Extracellular Glycolipids of Yeasts

Extracellular Glycolipids of Yeasts

Biodiversity, Biochemistry, and Prospects

Ekaterina Kulakovskaya

Tatiana Kulakovskaya

Skryabin Institute of Biochemistry and Physiology of Microorganisms, Russian Academy of Sciences

AMSTERDAM • BOSTON • HEIDELBERG • LONDON
NEW YORK • OXFORD • PARIS • SAN DIEGO
SAN FRANCISCO • SINGAPORE • SYDNEY • TOKYO

ELSEVIER

Academic Press is an imprint of Elsevier

Academic Press is an imprint of Elsevier
The Boulevard, Langford Lane, Kidlington, Oxford, OX5 1GB, UK
225 Wyman Street, Waltham, MA 02451, USA

First published 2014

British Library Cataloguing in Publication Data
A catalogue record for this book is available from the British Library

Library of Congress Cataloging-in-Publication Data
A catalog record for this book is available from the Library of Congress

ISBN: 978-0-12-420069-2

For information on all Academic Press publications
visit our website at store.elsevier.com

This book has been manufactured using Print On Demand technology. Each copy is produced to
order and is limited to black ink. The online version of this book will show color figures where
appropriate.

CONTENTS

ACKNOWLEDGMENTS

The experimental part of the work was done in the Skryabin Institute of Biochemistry and Physiology of Microorganisms, Russian Academy of Sciences.

Special thanks to Dr. W.I. Golubev, the discoverer of antifungal activity of cellobiose lipid producers, for providing yeast strains and for fruitful discussion. We are grateful to our colleagues Drs. E.O. Puchkov, A.S. Shashkov, N.E. Nifantiev, A. Zinin, Y. Tsvetkov, A. Grachev, and A. Ivanov for their great experimental contributions to and creative interpretations of the results. We thank Elena Makeeva for her help with preparing the manuscript. This study was supported by the Russian Foundation for Basic Research, Projects Nos. 06-04-48215, 06-04-08253-ofi, and 12-04-32138-mol-a.

INTRODUCTION

Microorganisms are characterized by a great diversity of the so-called secondary metabolites, that is, compounds that are not obligatory participants of metabolism but, nevertheless, provide advantages to producers in their survival under unfavorable environmental conditions and competition for ecological niches. Many of these compounds are biologically active and, hence, have good and promising applications in industry, agriculture, and medicine.

Secondary metabolites include the so-called biosurfactants: lipopeptides, glycolipids, fatty acids, neutral lipids, and phospholipids, as well as some amphiphilic biopolymers. These substances are widespread in microorganisms, from bacteria to fungi. They were found during the studies of microbial growth on hydrophobic substrates, including oils and hydrocarbons, and were supposed to improve the solubility and bioavailability of these substrates. The properties of biosurfactants of different chemical nature and origin, as well as their research and commercial prospects, have been described in a number of reviews (Lang and Wagner, 1987; Rosenberg and Ron, 1999; Kitamoto et al., 2002; Rodrigues et al., 2006; Langer et al., 2006; Arutchelvi et al., 2008; Van Bogaert et al., 2007, 2011). Many reviews are devoted to future potential of biosurfactants in medicine and industry (Banat et al., 2010; Fracchia et al., 2012; Marchant and Banat, 2012; Cortés-Sánchez et al., 2013). Springer Publishers have issued a volume "Biosurfactant" in the series "Advances in Experimental Medicine and Biology" (Sen, 2010) and a volume "Biosurfactants. From Genes to Applications" in the series "Microbiology Monographs" (Soberón-Chávez, 2011).

The following properties of these compounds make them relevant for life science and biotechnology:

- structural diversity;
- multiple biological activities;
- biodegradability;
- nontoxicity;

- the possibility of inexpensive production using simple nutrient media, including those containing industrial and agricultural wastes;
- promising applications as detergents, antibiotics, and amphiphilic compounds.

The extracellular glycolipids of yeast and fungi belong to biosurfactants. These compounds are glycosides of fatty acids containing one or more monosaccharide residues that may contain additional O-substituents at the sugar moiety.

These compounds are mentioned in many reviews on biosurfactants (Lang and Wagner, 1987; Rosenberg and Ron, 1999; Kitamoto et al., 2002; Cameotra and Makkar, 2004; Rodrigues et al., 2006; Langer et al., 2006). However, the reviews devoted specifically to yeast extracellular glycolipids are few (Van Bogaert et al., 2007a,b, 2011; Arutchelvi et al., 2008; Bölker et al., 2008; Kulakovskaya et al., 2008, 2009; Arutchelvi and Doble, 2011; Van Bogaert and Soetaert, 2011

The studies of yeast extracellular glycolipids attract attention due to their numerous activities: from biosurfactant properties providing utilization of hydrophobic substrates to fungicidal properties, as well as a number of other biological activities that make these compounds scientifically and practically promising.

Structural diversity, numerous biological activities, biodegradability, nontoxicity, and possibility of inexpensive production make them attractive for future applications in industry, cosmetology, medicine and agriculture as ecologically pure detergents, fungicides of new generation, and other useful products. Up to date, scientific literature has accumulated quite a lot of data on these compounds, which should be generalized for better understanding of the potential of yeast as a producer of biologically active substances, for development of ecological biotechnologies and research reagents. Although the biological role of extracellular glycolipids in nature is associated primarily with their surfactant properties, the detection of antifungal activity against a broad spectrum of yeast-like fungi in cellobiose lipids (representatives of these compounds) suggests that glycolipid secretion may play a key role in the adaptation to unfavorable environmental conditions. The study of structural peculiarities, the mechanism of action, and distribution of these natural fungicides may be important for a better understanding of antagonistic relationship between microorganisms, as well as the prospects of their practical application as compounds for plant and crop protection from phytopathogenic fungi and antibiotics and biologically active compounds in medicine.

Generalization of the data on the biochemistry, cell biology, and biotechnology of extracellular fungal glycolipids is of concern for microbiologists, biochemists, biotechnologists, and students of the respective specialties.

The book presents modern data on the yeasts producing extracellular glycolipids, their composition, structure and properties, biosynthetic pathways, methods of isolation and identification, antifungal activity, and mechanisms of action. The applied potential of these compounds in medicine, agriculture, and industry is being considered. The emphasis is placed on cellobiose lipids, including their structure, distribution, and antifungal activity.

CHAPTER 1

Structure and Occurrence of Yeast Extracellular Glycolipids

Secretion of glycolipids, namely fatty acid glycosides, was found in mycoplasms, bacteria (including actinobacteria), mixomycetes, fungi, plants, ascidia, and nematodes. The most-known extracellular glycolipids of yeast fungi are cellobiose lipids, mannosylerythritol lipids (MELs), and sophorolipids.

1.1 THE STRUCTURES OF EXTRACELLULAR GLYCOLIPIDS OF YEAST

1.1.1 Cellobiose Lipids

Cellobiose lipids consist of a residue of cellobiose, the disaccharide composed of two glucose residues linked by a $1,4'$-β-glycoside bond, and fatty acid residue as an aglycone.

The simplest compound of this group consists of a cellobiose residue linked through a glycosidic bond to 2,15,16-trihydroxyhexadecanoic acid (Figure 1.1A). The diversity of cellobiose lipids is determined by O-substituents in cellobiose residue and by the number of hydroxyl groups in fatty acid residue. The cellobiose residue may contain acetate groups and/or C_6 or C_8 hydroxy fatty acids as O-substituents (Figure 1.1B, C).

According to the terminology of the review (Kitamoto et al., 2002), the cellobiose lipid without O-substituents in the cellobiose residue is named cellobiose lipid A; those containing C_6 or C_8 hydroxy acids as O-substituents, as well as one or two acetate groups, are named cellobiose lipid B; and the methylated form is named cellobiose lipid C. This terminology has not become prevalent, and the authors of most publications either use the IUPAC nomenclature or call the compounds under study cellobiose lipids, adding the species name of the producer. Authors' trivial names may also be encountered: flocculosin for the cellobiose lipid of *Pseudozyma flocculosa* (Mimee et al., 2005), although such compound is found as a minor in *Ustilago maydis* (Kitamoto et al., 2002; Bolker et al., 2008).

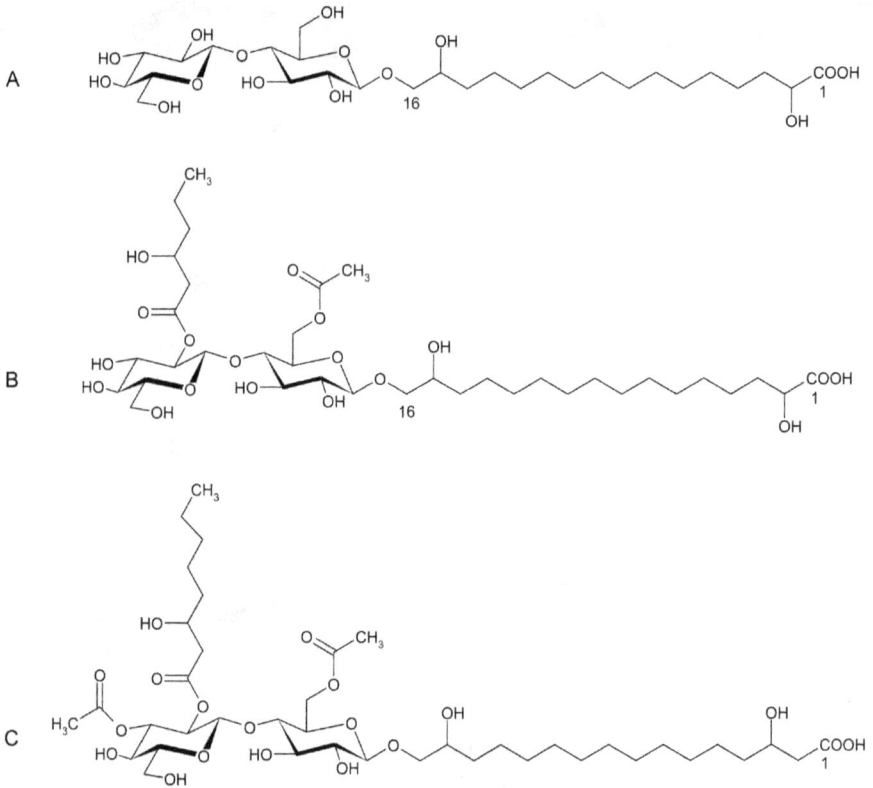

Figure 1.1 Cellobiose lipids of (A) Sympodiomycopsis paphiopedili, (B) Pseudozyma fusiformata, and (C) Pseudozyma flocculosa.

Extracellular cellobiose lipids were isolated for the first time from the culture liquid of smut fungus *U. maydis* (*zeae*) and named ustilagic acids in accordance with the generic name of the producer (Haskins and Thorn, 1951; Lemieux, 1951; Lemieux et al., 1951).

U. maydis was shown to secrete a mixture of non-acylated and acylated derivatives of β-D-cellobiosyl-2,16-dihydroxyl hexadecanoic acid and β-D-cellobiosyl-2,15,16-trihydroxyl hexadecanoic acid, including a relatively rare cellobiose lipid, methylated by the carboxylic group of 2,15,16-trihydroxyhexadecanoic acid (Frautz et al., 1986; Spoeckner et al., 1999; Bolker et al., 2008).

In *Pseudozyma fusiformata* (Kulakovskaya et al., 2005) and *Pseudozyma graminicola* (Golubev et al., 2008b), the major secreted

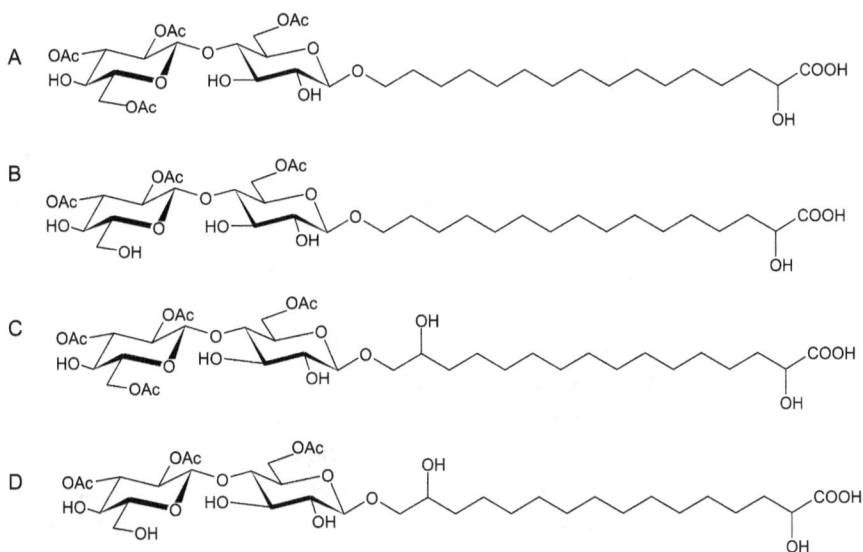

Figure 1.2 Structure of (A) major and (B–D) minor glycolipids of Cryptococcus humicola and Trichosporon porosum.

glycolipid is 2-O-3-hydroxyhexanoyl-β-D-glucopyranosyl-(1→4)-6-O-acetyl-β-D-glucopyranosyl-(1→16)-2,15,16-trihydroxyhexadecanoic acid (Figure 1.1B); however, some strains of *Ps. fusiformata* also secrete a simpler cellobiose lipid, having no 3-hydroxyhexanoic acid residue as an O-substituent.

The major extracellular glycolipid of the yeasts *Cryptococcus humicola* (Puchkov et al., 2002) and *Trichosporon porosum* (Kulakovskaya et al., 2010) is 2,3,4-O-triacetyl-β-D-glucopyranosyl-(1→4)-6-O-acetyl-β-D-glucopyranosyl-(1→16)-2,16-dihydroxyhexadecanoic acid (Figure 1.2A). Minor glycolipids of *Cr. humicola* were revealed containing C_{18} fatty acids with additional hydroxyl groups (Puchkov et al., 2002). Cellobiose lipids differing in the degree of acetylation and in the number of hydroxyl groups in the fatty acid residue were also obtained as minor components from the culture liquid of *Cr. humicola* strains (Puchkov et al., 2002; Kulakovskaya et al., 2006) and *T. porosum* (Kulakovskaya et al., 2010) (Figure 1.2B–D). The differences in cellobiose lipid composition of several strains of *Cr. humicola* were associated with prevalence of compounds with the four or three acetate groups in cellobiose residues (Kulakovskaya et al., 2006).

1.1.2 Mannosylerythritol Lipids

The structural peculiarities of MELs are described in a number of reviews (Kitamoto et al., 2002; Arutchelvi et al., 2008; Morita et al., 2009a; Arutchelvi and Doble, 2011). These glycolipids consist of a mannose residue etherified by erythrite at position 1. One or two fatty acid residues with a number of carbon atoms from 4 to 12 may be present in the mannose residue as O-substituents. The MELs are subdivided into three groups: MEL-A, MEL-B, and MEL-C, different in the quantity and position of acetate groups as O-substituents in the mannose residue (Kitamoto et al., 2002; Arutchelvi et al., 2008; Morita et al., 2009a; Arutchelvi and Doble, 2011) (Figure 1.3). Each of these groups includes a set of glycolipids which differ in the number of fatty acid residues as O-substituents in the mannose residue (mono- and diacylated MELs). Triacylated MELs etherified by the fatty acid residue at the terminal hydroxyl group of erythrite have been found in some strains of *Pseudozyma antarctica* and *Pseudozyma rugulosa* (Fukuoka et al., 2007a). In addition, there may also be numerous MEL stereoisomers.

MELs were found first as minor oily components in culture suspension of *U. maydis* (Haskins et al., 1955; Fluharty and O'Brien, 1969). MEL of *Ustilago* was characterized as a mixture of partially acylated derivatives of 4-O-β-D-mannopyranosyl-D-erythritol containing C_2, C_{12}, C_{14}, C_{16}, and C_{18} fatty acids residues (Bhattacharjee et al., 1970). MELs are major extracellular glycolipids of many species belonging to *Pseudozyma* genera (Kitamoto et al., 1990a,b, 1992a,b, 1993, 1995, 1998, 1999, 2001a; Fukuoka et al., 2007a,b, 2008a,b, 2012; Morita et al., 2006a,b, 2007, 2008a−d, 2009a,b, 2010a, 2011c, 2012, 2013). It has been shown that some or other MEL variants may be dominant in certain producers (Table 1.1). Most of the producers secrete not individual compounds but whole sets of MELs with different degrees of acylation and chain lengths of fatty acid residues.

The following rarely-occurring extracellular mannose-containing glycolipids have been found in *Pseudozyma parantarctica*: mannosylribitol lipids (with ribitol instead of erythrite), mannosylarabitol lipids (with arabitol instead of erythrite), and mannosylmannitol lipids (with mannitol instead of erythrite) (Morita et al., 2009a, 2012). *Pseudozyma shanxiensis* was found to produce more hydrophilic glycolipids than the previously-reported MELs. These MELs possessed a much shorter

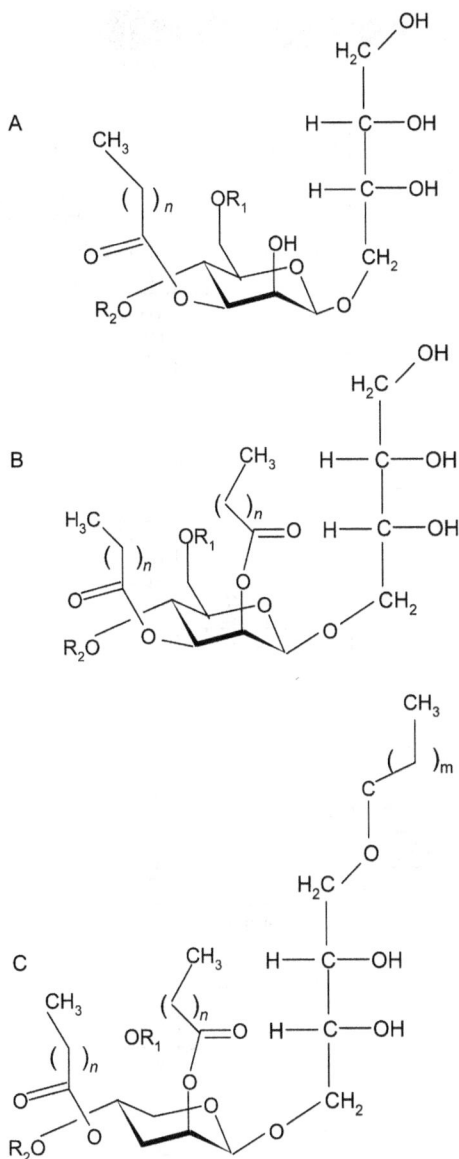

Figure 1.3 Structures of MELs: (A) monoacylated MEL, (B) diacylated MEL, and (C) triacylated MEL; MEL-A: R1 = Ac, R2 = Ac; MEL-B: R1 = Ac, R2 = H; MEL-C: R1 = H, R2 = Ac; n = 4–12; m = 6–16.

chain C-2 or C-4 at the C-2′ position of the mannose moiety compared to the MELs hitherto reported, which mainly possess a medium-chain acid C-10 at the position (Fukuoka et al., 2007b). *Pseudozyma chura-shimaensis* sp. was now found to produce a mixture of MELs,

Table 1.1 The Major Extracellular Glycolipids of Yeast Fungi and Their Producers

IUPAC (or Trivial) Names	Species	References
Cellobiose Lipids		
β-D-Glucopyranosyl-(1→4)-β-D-glucopyranosyl-(1→16)-2,15,16-trihydroxyhexodecanoic acid	*Ustilago maydis*	Haskins and Thorn (1951), Lemieux (1951), Lemieux et al. (1951), Bhattacharjee et al. (1970), Frautz et al. (1986), Spoeckner et al. (1999)
	Sympodiomycopsis paphiopedili	Golubev et al. (2004), Kulakovskaya et al. (2004)
2-o-3-Hydroxyhexanoil-β-D-glucopyranosyl-(1→4)-6-o-acetyl-β-D-glucopyranosyl-(1→16)-2,15,16-trihydroxyhexodecanoic acid	*Ustilago maydis*	Haskins and Thorn (1951), Lemieux (1951), Lemieux et al. (1951), Bhattacharjee et al. (1970), Frautz et al. (1986), Spoeckner et al. (1999)
	Pseudozyma fusiformata	Kulakovskaya et al. (2005, 2007)
	Pseudozyma graminicola	Golubev et al. (2008a,b)
2-o-3-Hydroxyoctanoil-3-o-acetyl-β-D-glucopyranosyl-(1→4)-6-o-acetyl-β-D-glucopyranosyl-(1→16)-3,15,16-trihydroxyhexodecanoic acid	*Ustilago maydis*	Haskins and Thorn (1951), Lemieux (1951), Lemieux et al. (1951), Bhattacharjee et al. (1970), Frautz et al. (1986), Spoeckner et al. (1999)
	Pseudozyma flocculosa	Mimee et al. (2005)
2,3,4-o-Triacetyl-β-D-glucopyranosyl-(1→4)-6-o-acetyl-β-D-glucopyranosyl-(1→16)-2,16-dihydroxyhexodecanoic acid	*Cryptococcus humicola*	Puchkov et al. (2002), Kulakovskaya et al. (2006, 2007), Morita et al. (2011a), Imura et al. (2012)
	Trichosporon porosum	Kulakovskaya et al. (2010)
Mannosylerythritol Lipids (MELs)		
MEL-A		
4-O-[(4′,6′-di-O-acetyl-3′-O-alkanoil)-β-D-mannopyranosil] *meso*-erythritol	*Ustilago maydis*	Fluharty and O'Brien (1969), Spoeckner et al. (1999), Kurz et al. (2003)
4-O-[(4′,6′-di-O-acetyl-2′,3′-di-O-alkanoil)-β-D-mannopyranosyl] *meso*-erythritol	*Pseudozyma crassa*	Fukuoka et al. (2008a)
4-O-[(4′,6′-di-O-acetyl-2′,3′-di-O-alkanoil)-β-D-mannopyranosyl] *meso*-erythritol-alkanoil	*Pseudozyma antarctica*	Kitamoto et al. (1990a,b, 1992a,b, 1999), Morita et al. (2007), Fukuoka et al. (2007a)
	Pseudozyma aphidis	Rau et al. (2005)
	Pseudozyma churashimaensis	Morita et al. (2011c)
	Pseudozyma parantarctica	Morita et al. (2007, 2008c, 2012)
	Pseudozyma rugulosa	Morita et al. (2006a)

(Continued)

Table 1.1 (Continued)

IUPAC (or Trivial) Names	Species	References
	Pseudozyma fusiformata	Morita et al. (2007) Konishi et al. (2007)
	Kurtzmanomyces sp.	Kakugawa et al. (2002)
MEL-B		
4-*O*-[(6′-*O*-acetyl-3′-*O*-alkanoil)-β-D-mannopyranosyl] *meso*-erythritol	*Ustilago maydis*	Fluharty and O'Brien (1969), Spoeckner et al. (1999), Kurz et al. (2003)
4-*O*-[(6′-*O*-acetyl-2′,3′-di-*O*-alkanoil)-β-D-mannopyranosyl] *meso*-erythritol	*Ustilago scitaminea*	Morita et al. (2011b)
	Pseudozyma churashimaensis	Morita et al. (2011c)
4-*O*-[(6′-*O*-acetyl-2′,3′-di-*O*-alkanoil)-β-D-mannopyranosyl] *meso*-erythritol-alkanoil	*Pseudozyma crassa*	Fukuoka et al. (2008a)
	Pseudozyma tsukubaensis	Fukuoka et al. (2008b)
	Pseudozyma antarctica	Kitamoto et al. (1990a,b, 1992a,b, 1999), Morita et al. (2007), Fukuoka et al. (2007a)
	Kurtzmanomyces sp.	Kakugawa et al. (2002)
MEL-C		
4-*O*-[(6′-*O*-acetyl-2′,3′-di-*O*-alkanoil)-β-D-mannopyranosyl] *meso*-erythritol-alkanoil4-*O*-[(4′-*O*-acetyl-3′-*O*-alkanoil)-β-D-mannopyranosyl] *meso*-erythritol	*Ustilago maydis*	Fluharty and O'Brien (1969), Spoeckner et al. (1999), Kurz et al. (2003)
4-*O*-[(6′-*O*-acetyl-2′,3′-di-*O*-alkanoil)-β-D-mannopyranosyl] *meso*-erythritol-alkanoil4-*O*-[(4′-*O*-acetyl-2′,3′-di-*O*-alkanoil)-β-D-mannopyranosyl] *meso*-erythritol	*Ustilago cynodontis*	Morita et al. (2008a)
4-*O*-[(6′-*O*-acetyl-2′,3′-di-*O*-alkanoil)-β-D-mannopyranosyl] *meso*-erythritol-alkanoil4-*O*-[(4′-*O*-acetyl-2′,3′-di-*O*-alkanoil)-β-D-mannopyranosyl] *meso*-erythritol-alkanoil	*Pseudozyma churashimaensis*	Morita et al. (2011c)
	Pseudozyma crassa	Fukuoka et al. (2008a)
	Pseudozyma antarctica	Kitamoto et al. (1990a,b, 1992a,b, 1999), Morita et al. (2007), Fukuoka et al. (2007a)
	Pseudozyma graminicola	Morita et al. (2008d)
	Pseudozyma hubeiensis	Konishi et al. (2007, 2011)
	Pseudozyma shanxiensis	Fukuoka et al. (2007b)
	Pseudozyma siamensis	Morita et al. (2008b)
	Kurtzmanomyces sp.	Kakugawa et al. (2002)

(Continued)

Table 1.1 (Continued)

IUPAC (or Trivial) Names	Species	References
Other Mannose Lipids		
Mannosylmannitol lipids	*Pseudozyma parantarctica*	Morita et al. (2012)
Mannosylribitol lipids	*Pseudozyma parantarctica*	Morita et al. (2012)
Mannosylarabitol lipids	*Pseudozyma parantarctica*	Morita et al. (2012)
Sophorolipids		
6-*o*-acetyl-β-D-glucopyranosyl-(1→2)-(6′-*o*-acetyl-β-D-glucopyranosyl)-17-hydroxy-octadecenoic acid	*Starmerella bombicola*	Ito and Inoue (1982), Rau et al. (1996), Daniel et al. (1998, 1999), Casas and Garcia-Ochoa (1999), Pekin et al. (2005), Kurtzman et al. (2010), Van Bogaert et al. (2010), Takahashi et al. (2011), Gupta and Prabhune (2012)
	Wickerhamiella domercqiae	Chen et al. (2006a,b), Ma et al. (2011, 2012), Li et al. (2012)
6-*o*-acetyl-β-D-glucopyranosyl-(1→2)-(6′-*o*-acetyl-β-D-glucopyranosyl)-18-hydroxy-octadecenoic acid	*Candida batistae*	Konishi et al. (2008)
6-*o*-acetyl-β-D-glucopyranosyl-(1→2)-(6-*o*-acetyl-β-D-glucopyranosy)-13-hydroxydocosanoic acid	*Rhodotorula bogoriensis*	Tulloch et al. (1968), Cutler and Light (1979), Zhang et al. (2011)
	Candida apicola	Gorin et al. (1960), Tulloch and Spencer (1966), Hommel et al. (1994)

including a novel tri-acetylated derivative MEL-A2 (Morita et al., 2011c). The MEL-B comprising a hydroxy fatty acid was revealed under study of MEL production of *Pseudozyma tsukubaensis*: 1-*O*-β-(2′-*O*-alka(e)noyl-3′-*O*-hydroxyalka(e)noyl-6′-*O*-acetyl-D-mannopyranosyl)-D-erythritol (Yamamoto et al., 2013).

1.1.3 Sophorolipids

Sophorolipid comprise a residue of sophorose, the disaccharide consisting of two glucose residues linked by the β-1,2′ bond, and fatty acid as an aglycone (Figure 1.4). It can be acetylated on the 6- and/or 6′-positions of sophorose residue. One terminal or subterminal hydroxylated fatty acid is β-glycosidically linked to the sophorose molecule. The hydroxy fatty acid residue can have one or more unsaturated bonds (Figure 1.4). The carboxylic group of fatty acid is either free (acidic or open form) or internally esterified (lactonic form) (Figure 1.5).

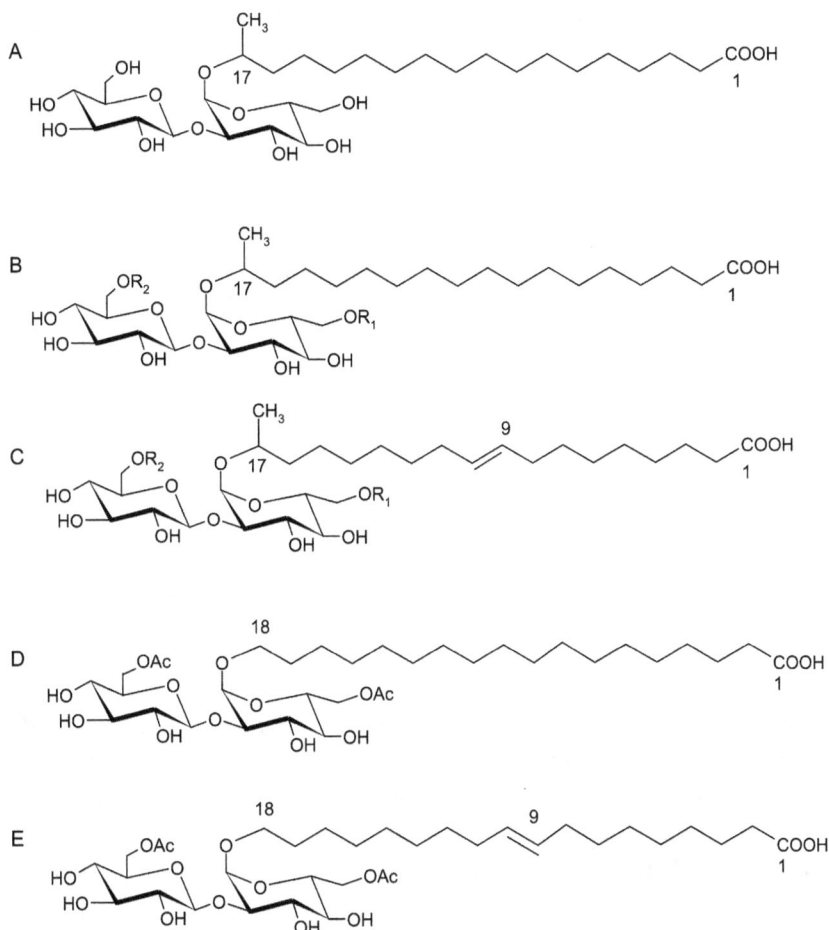

Figure 1.4 Structures of sophorolipids in acid form: (A) deacetylated sophorolipid, (B, C) major sophorolipids of Starmerella bombicola, and (D, E) major sophorolipids of Candida batistae.

Sophorolipids can exist in the form lactones both in monomeric or in dimeric forms (Nunez et al., 2004).

Such glycolipids containing C_{22} fatty acid residue were found for the first time in *Torulopsis magnoliae* (*Candida magnolia, Candida apicola*) (Gorin et al., 1961; Tulloch and Spencer, 1966). *Candida bombicola* (*Starmerella bombicola*) is currently the well-studied producer of sophorolipids.

The structures of sophorolipids from different yeast species are described in detail in reviews (Van Bogaert et al., 2007, 2011). The

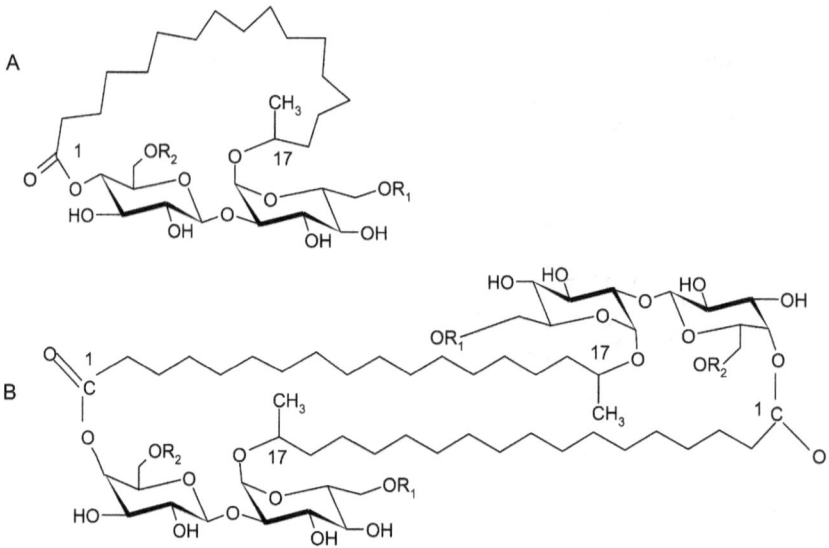

Figure 1.5 Structures of sophorolipids lipids in lactone form: (A) monomeric lactone and (B) dimeric lactone.

main producers are listed in Table 1.1. Sophorolipids differ in the number and position of acetate groups as O-substituents in the carbohydrate reside and in the structures of fatty acid residues (Figure 1.4). For example, sophorolipids of *St. bombicola* and *Candida batistae* differ in the position of hydroxylic group in fatty acid residue: the fatty acid residues in sophorolipids of *St. bombicola* are hydroxylated mainly in $\omega - 1$ position, while that of *C. batistae* are hydroxylated mainly in ω-position (Konishi et al., 2008) (Figure 1.4).

The glycolipid produced by *Rhodotorula bogoriensis* contains C_{22} fatty acid residue as an aglycone (Tulloch et al., 1968; Nunez et al., 2004) (Figure 1.6).

1.2 GLYCOLIPID OCCURRENCE IN EUMYCETES

Extracellular glycolipids were found in eumycetes, mainly in yeast or yeast-like fungi. Filamentous fungi are mentioned only in single reports.

The so-called roselipins (Figure 1.7) (consisting of C_{20}-fatty acids with three hydroxyl groups, mannose, and arabitol residues) (Tabata et al., 1999) are synthesized by *Clonostachys rosea* (= *Gliocladium roseum*), which is an anamorpha of the ascomycete *Bionectria ochroleuca*.

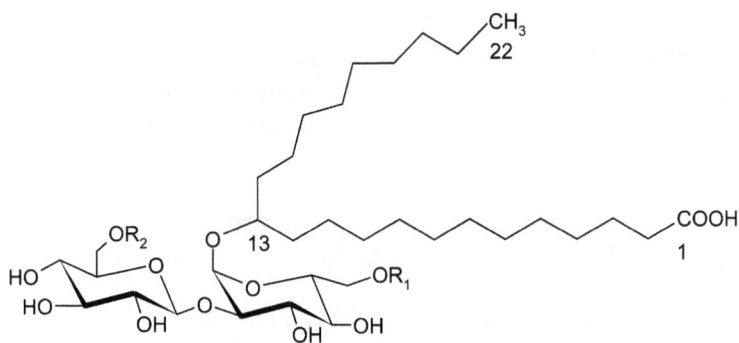

Figure 1.6 Structure of major sophorolipid of Rhodotorula bogoriensis (Tulloch et al., 1968).

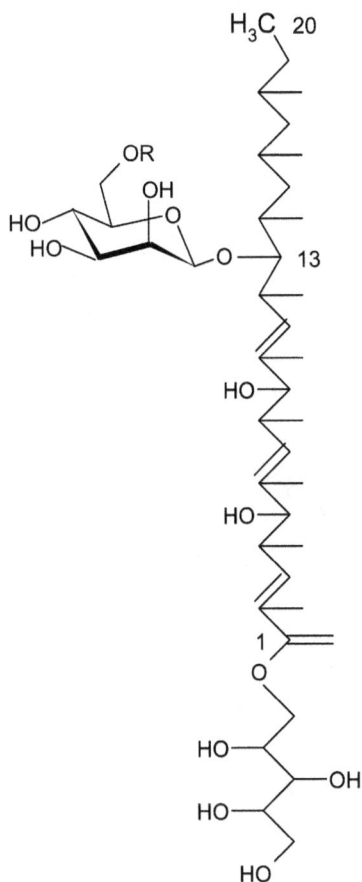

Figure 1.7 Structure of roselipin (Tabata et al., 1999).

Monoglycosyloxydecenic acid was found in *Aspergillus niger* (Laine et al., 1972). The glycolipids comprising glucose and galactose residues, oxalate, and 17-hydroxydocosanoic acid (emmyguyacins, Figure 1.8) were isolated from an unidentified fungus (Boros et al., 2002).

The fungus *Dacryopinus spathularia* produces rare glycolipids (Stadler et al., 2012). One of them is shown in Figure 1.9.

For yeasts, the biosynthesis of extracellular glycolipids is characteristic of certain taxa. In particular, the glycolipids containing sophorose are produced mostly by ascoporous yeasts (class Saccharomycetes, order Saccharomycetales) of the genera *Starmerella* (Kurtzman et al., 2010), *Wickerhamiella* (Chen et al., 2006a,b), *Wickerhamomyces* (Thaniyavarn et al., 2008), and the phylogenetically related asporogenous species of the genus *Candida* (Price et al., 2012). Several sophorolipid producers belonging to *Starmerella* clade were identified: *Candida*

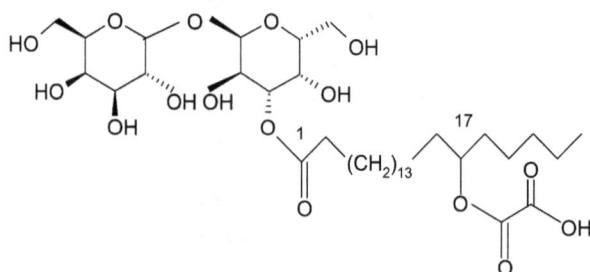

Figure 1.8 Structure of emmyguyacin (Boros et al., 2002).

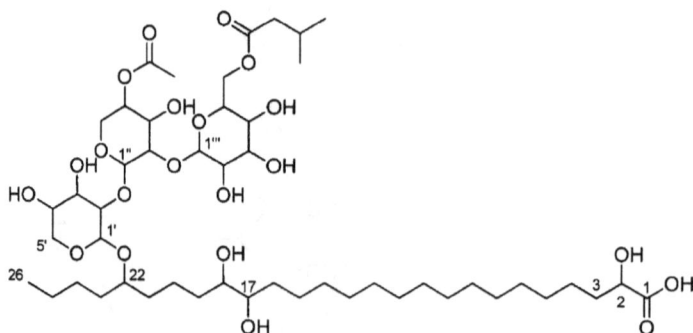

Figure 1.9 Structure of a representative of glycolipids of Dacryopinus spathularia (Stadler et al., 2012).

riodocensis, *Candida stellata* (Kurtzman et al., 2010), and *Candida floricola* (Imura et al., 2010).

The only exception among sophorolipid-forming yeasts is *Rh. bogoriensis* (Tulloch et al., 1968), which is phlogenetically related to basidiomycetes of the class Microbotryomycetes.

On the contrary, the glycolipids containing cellobiose are synthesized almost exclusively by basidiomycetes, mainly members of the order Ustilaginales (class Ustilaginomycetes): the species of the genera *Pseudozyma* (Golubev et al., 2001) and *Ustilago* (Haskins, 1950). Individual producers of cellobiose lipids in basidiomycetes were also found in the classes Exobasidiomycetes and Tremellomycetes. In the former, this is the species *Sympodiomycopsis paphiopedili* (order Microstromatales) (Golubev et al., 2004); in the latter, these are the species of the order Trichosporonales, the genera *Cryptococcus* (Puchkov et al., 2001) and *Trichosporon* (Kulakovskaya et al., 2010).

The cellobiose lipid-producing species of the above genera often secrete MELs. These compounds are especially widespread among *Pseudozyma* clade (Kitamoto et al., 1990a,b, 1992a,b; Fukuoka et al., 2008a,b; Morita et al., 2007, 2008b−d, 2012; Konishi et al., 2007).

Schizonella also related to Ustilaginales (Deml et al., 1980) and *Kurzmanomyces* (order Agaricostilbales, class Agaricostilbomycetes) can also be added to the above genera (Kakugawa et al., 2002).

Due to the development of fungal systematics, species are quite often redefined and their names given in any previous works should be critically considered. The most-studied eumycetes producing extracellular glycolipids are defined in Table 1.1.

CHAPTER 2

Methods for Studying Yeast Extracellular Glycolipids

2.1 CULTURE MEDIA AND METHODS FOR INCREASING THE YIELD OF YEAST EXTRACELLULAR GLYCOLIPIDS

The basic principles for selecting the nutrient media to obtain fungal extracellular glycolipids are as follows:

- *Excess of carbon sources*: These may be sugars and fatty acids as well as hydrocarbons or their combinations for some species. The addition of a considerable excess of glucose (up to 10% and more) to the medium, when the stationary phase has been reached after the growth at glucose content of 1–2%, is an effective technique. High sugar concentrations inhibit the growth of many fungi and, therefore, it is not always expedient to add them at the beginning of cultivation.
- *Nitrogen starvation*: It is important for some species but is not a decisive factor for other species.
- *Intensified aeration*: is required for utilization of hydrocarbons and fatty acids as carbon sources, while under sugar consumption the cultivation without stirring can be used.

Extracellular glycolipids can be obtained by cultivation both in flasks and in fermenters; chemostat cultures are often used in the latter case. It is obvious that different producers yield different amounts of target products in the same media, and the optimization of production of each particular glycolipid remains a nontrivial problem. Here, we will consider the particular examples of how these approaches are implemented.

Enhanced production of some extracellular glycolipids was observed in the media with hydrophobic carbon sources, including carbohydrates and fats. This approach is effective for bacterial rhamnolipid, mannosylerythritol lipids, and sophorolipids (Kitamoto et al., 2002). It may be due to both by the use of fatty acids taken up from the medium for the synthesis of these compounds and by the fact that these extracellular glycolipids are needed as detergents for solubilization and consumption of fatty acid substrates and, hence, their biosynthesis may be an induced process.

The comparison of bacterial producers of biosurfactants (including rhamnolipids) with fungal producers demonstrates the higher productivity of fungi, especially in relatively simple media. So, the best productivity for the rhamnolipid producer *Pseudomonas* sp. was 45 g/l (Muthusamy et al., 2008), while that for the sophorolipid-producing yeast was about 400 g/l (Pekin et al., 2005).

The Appendix presents several variants of relatively simple nutrient media and cultivation methods for obtaining extracellular glycolipids under laboratory conditions.

2.1.1 Cellobiose Lipids

In the initial stage of research, extracellular glycolipids of yeast fungi were obtained using the conventional media containing glucose as a carbon source, yeast extract, and mineral salts. However, the level of production of these compounds was low. In particular, we obtained 13−50 mg/l of cellobiose lipids from *Cr. humicola* and *Pseudozyma* sp.

The yield of cellobiose lipids was increased by use of fats as a carbon source: *U. maydis* produced 16 g/l of cellobiose lipids when grown on the media with coconut oil (Frautz et al., 1986). The media with 30−50 g/l glucose or 50 g/l saccharose, 1.7 g/l yeast nitrogen base (without amino acids and $(NH_4)_2SO_4$), and 1.3 g/l $(NH_4)_2SO_4$ were used for obtaining cellobiose lipid of *U. maydis* (Gűnter et al., 2010). After cultivation at 30°C and 120 rpm for 7−10 days, the yield of cellobiose lipid was 16−20 g/l. The optimal pH was 3−3.5 (Gűnter et al., 2010).

In contrast to *U. maydis*, nitrogen starvation did not enhance the cellobiose lipid production by *Cr. humicola* (Morita et al., 2011a). The authors used a technique consisting of the initial biomass production under stirring for several days, followed by the addition of excess glucose up to 10%, to obtain glycolipids during a long-term cultivation. Cellobiose lipid production by *Cr. humicola* was 13.1 g/l (Morita et al., 2011a).

The factors and conditions that affected the production of the antifungal glycolipid flocculosin by *Ps. flocculosa* (Hammami et al., 2008) were studied. Concentration of the start-up inoculum was found to play an important role in flocculosin production, as the optimal level increased the productivity by as much as 10-fold. If conditions were conducive for the production of the glycolipid, carbon availability appeared to be the only limiting factor. Inorganic nitrogen starvation did not trigger production of flocculosin (Hammami et al., 2008).

2.1.2 Mannosylerythritol Lipid

The productivity of various yeast species and optimization of the yield of MEL have been investigated. Rather high yields of MEL were obtained: *Ps. antarctica* produced more than 40 g/l of MEL on the media containing oleic acid, glycerol, and soybean oil (Kitamoto et al., 1992a,b; Kim et al., 1999) and even up to 140 g/l in the media with *n*-octodecane (Kitamoto et al., 2001a). For *Ustilago scitaminea*, the optimal medium for MEL-B production (25.1 g/l) contained sugarcane juice (19% sugars) and 1 g/l of urea (Morita et al., 2011b). The yeast *Ps. rugulosa* produced MEL-A (68%), MEL-B (12%), and MEL-C (20%) (Morita et al., 2006a,b). During the cultivation under stirring on the media with soybean oil as a carbon source and sodium nitrate as a nitrogen source, the total yield of MEL was up to 142 g/l (Morita et al., 2006a,b). *Ps. antarctica* and *Ps. parantarctica* yielded up to 30 g/l during 7 days of cultivation in the simplest medium containing 8% soybean oil, 0.3% $NaNO_3$, 0.03% $MgSO_4$, 0.03% KH_2PO_4, and 0.1% yeast extract (pH 6.0) (Morita et al., 2007). *Pseudozyma aphidis*, *Ps. rugulosa*, and *Ps. tsukubaensis* were a little inferior to them (about 25 g/l). *Ps. fusiformata* yielded less than 5 g/l in the same medium. *Ps. aphidis* produced MEL when cultivated on glucose; the addition of mannose and erythritol as extra carbon sources increased glycolipid production. Fractional addition of soybean oil by 20 ml/l up to the total concentration of 80 ml/l resulted in obtaining up to 75 g/l of glycolipids during 10 days under stirring (Rau et al., 2005). *Pseudozyma crassa* produced 4.6 g/l of MEL in the medium with glucose and oleic acid (Fukuoka et al., 2008a).

Ps. parantarctica JCM 11752 produced quite a lot of mannosyl mannitol lipids: 18.2 g/l (Morita et al., 2009a).

For MEL production, *U. maydis* was grown in a medium containing 1% yeast extract, 2% peptone, and 2% sucrose, and then exposed to nitrogen starvation in a medium containing 5% sucrose, vitamins, and trace elements (Hewald et al., 2006).

As has been shown for *Ps. antarctica* and *Pseudozyma apicola*, the synthesis of glycolipids increases 7–8.5-fold when the medium is enriched in food and fragrance industry wastes: the fatty acid fraction obtained after plant oil refinement or soap production wastes containing a lot of fatty acids (Bednarski et al., 2004). Glycolipid production was 7.3–13.4 g/l and 6.6–10.5 g/l in the media with the addition of soap industry and plant oil refinement wastes, respectively.

2.1.3 Sophorolipids

Most of the work on enhancement of the yield of target glycolipids involve sophorolipids as they were the first extracellular yeast glycolipids that found a practical application. It has been shown that sophorolipids are effectively produced in media containing plant oils, glucose, or hydrocarbons (Tulloch et al., 1968; Cooper and Paddock, 1983; Hommel et al., 1994; Zhou and Kosaric, 1995; Rau et al., 1996; Davila et al., 1997; Casas and Garcia-Ochoa, 1999). These glycolipids are most effectively synthesized in the nitrogen-limited media with excessive carbon source (Daniel et al., 1999; Otto et al., 1999). As glycolipid molecules contain a fatty acid residue, the media with the higher content of fatty acids were used to increase their production. The yield is increased through controlled cultivation in fermenters (Kim et al., 2009).

A medium containing 3% glucose, 0.15% yeast extract, and tap water was used to obtain the sophorolipid of *Rh. bogoriensis* (Cutler and Light, 1979). Up to 5 g/l of the sophorolipid could be produced in such a medium. Sophorolipid production increased to ~20 g/l, if the content of glucose in the same medium was increased to 5–7.5%, but decreased five times if the content of yeast extract was increased to 2.4% (Cutler and Light, 1979).

Some works give lower values for *Rh. bogorensis*, probably due to the peculiarities of cultivation conditions. In the work of Zhang et al. (2011), this yeast produced only 0.33 g/l of sophorolipids in the medium with glucose and 1.26 g/l on the addition of rapeseed oil.

It was shown for the relatively-little-studied sophorolipid producer *Wickerhamiella domercqiae* that ammonium salts inhibited the synthesis of sophorolipids, while organic nitrogen increased their yield, especially in the lactone form (Ma et al., 2012). The yield of glycolipids increased in the case of alkalization of the medium (Ma et al., 2012).

Mutants capable of producing more sophorolipids than the parent strain were obtained for some producer species. The mutant strain of *W. domercqiae* yielded twice as much sophorolipids (104 g/l in flasks and 135 g/l in fermenter) than the parent strain (Li et al., 2012).

Different species and strains produce different quantities of sophorolipids in the same media. About 6 g/l and 20 g/l of the *C. batistae* CBS 8550 and *St. bombicola* NBRC 10243 sophorolipids, respectively, were obtained during 3-day cultivation in the medium (glucose, 50 g/l; olive oil, 50 g/l; $NaNO_3$, 3 g/l; KH_2PO_4, 0.5 g/l; $MgSO_4 \cdot 7H_2O$, 0.5 g/l; yeast extract, 1–5 g/l (pH 6.0)) in flasks under stirring (250 rpm) (Konishi et al., 2008).

For the time being, *St. bombicola* is a record holder in sophorolipid production. The basic principle of productivity enhancement is to use the nutrient media with excessive carbon sources, which are supplemented with hydrophobic substrates, primarily plant oils, and deficient in nitrogen sources. Mineral nitrogen sources are more preferable than organic ones.

St. bombicola produced more than 30 g/l of sophorolipids in a medium containing glucose and sunflower oil (Ito and Inoue, 1982). During 8 days of cultivation in the medium with 10% glucose, 10% sunflower oil, and 0.1% yeast extract, *St. bombicola* produced 120 g/l of sophorolipids (Casas and Garcia-Ochoa, 1999). The yield of sophorolipids increased to 420 g/l in a medium containing serum and rapeseed oil (Daniel et al., 1998).

Several hydrophilic carbon sources, hydrophobic cosubstrates, and nitrogen sources were supplied to culture media, and their influence on sophorolipid production in *St. bombicola* was evaluated (Ribeiro et al., 2013). The production of acidic $C_{18:1}$ diacetylated hydroxy fatty acid sophorolipid was favored when the culture media was supplied with avocado, argan, sweet almond, and jojoba oil or when $NaNO_3$ was supplied instead of urea. A lactonic $C_{18:3}$ hydroxy fatty acid diacetylated sophorolipid was detected when borage and onagra oils were used as cosubstrates (Ribeiro et al., 2013). To achieve high time–space efficiency for sophorolipid production with yeast *St. bombicola*, a strategy of high cell density fermentation was employed (Gao et al., 2013).

The cell density up to 80 g dry cell weight/l was obtained and a high productivity was achieved (>200 g/l per day). This productivity was attained on 24 h of cultivation, highlighting the industrial potential of this cultivation method (Gao et al., 2013).

2.1.4 Yeast Glycolipid Production in Low-Cost Media

Commercial application of yeast glycolipids requires the development of nutrient media based on inexpensive raw materials. Media on the basis of food industry wastes have been proposed for sophorolipid production.

St. bombicola was able to synthesize sophorolipids when cultivated in the media containing fatty acid esters obtained as a result of esterification with methanol (Ashby et al., 2006, 2010), as well as plant oil purification wastes (Bednarski et al., 2004). During the cultivation on molasses from sugarcane, *St. bombicola* produced 14.4 g/l and 22.8 g/l of sophorolipids in flasks and in fermenter, respectively, with the pH optimum of 6.0 (Takahashi et al., 2012). Glycerol-containing wastes of biodiesel production have to be utilized, which increases the cost of this fuel. The application of glycerol-containing wastes allows the production of relatively inexpensive sophorolipids and is useful for biodiesel cost reduction and market development. Model experiments showed that *St. bombicola* produced 6.6 g/l of sophorolipids in media containing 15% glycerol and 10% sunflower oil (Wadekar et al., 2012). The level of sophorolipid production was up to 60 g/l during the cultivation of *St. bombicola* on biodiesel coproduct stream containing up to 40% glycerol (Ashby et al., 2005).

The kinetics of growth of *St. bombicola*, sophorolipid production, and properties of sophorolipids were studied under cultivation in a cheap fermentative medium containing sugarcane molasses, yeast extract, urea, and soybean oil (Daverey Pakshirajan, 2009, 2010).

Table 2.1 shows the yields of fungal glycolipids in the low-cost media.

2.2 PURIFICATION METHODS

One of the widely-used approaches to the purification of extracellular glycolipids is the extraction by various organic solvents. The culture liquid is treated with ethyl acetate, and glycolipids move into the organic phase. In some works (including ours), culture liquids are

Table 2.1 The Yield of Yeast Glycolipids in Inexpensive Nutrient Media on the Basis of Plant Raw Materials

Raw Material	Glycolipid	Producer	Yield, g/l	References
Palm oil	Sophorolipids	*Candida lipolityca* IA 1055	11.72	Vance-Harrop et al. (2003)
Corn oil	Sophorolipids	*Starmerella bombicola* ATCC 22214	400	Pekin et al. (2005)
Glycerol-containing wastes from biodiesel production	Sophorolipids	*Starmerella bombicola*	60	Ashby et al. (2005)
Sugarcane molasses, yeast extract, urea, soybean oil	Sophorolipids	*Starmerella bombicola*	67.3	Daverey and Pakshirajan (2009)
Soybean oil	MELs	*Pseudozyma sp.* SY16	95	Kim et al. (2006)
Oil refinement waste	MELs	*Pseudozyma antarctica*	10.5	Deshpande and Daniels (1995)
Soybean oil	MELs	*Pseudozyma rugulosa*	142	Morita et al. (2006a)
Sugarcane juice	MEL-B	*Ustilago scintaminea* NBRC 32730	25.1	Morita et al. (2011b)

lyophilized, followed by methanol extraction of the target products from lyophilisate.

Characteristics of cellobiose lipids and MEL such as low solubility in aqueous solutions at pH below 2 is also used. In this case, precipitation is performed without removing the biomass. The yeast culture is acidified to pH values below 2 and glycolipids are adsorbed on the biomass, separated by centrifugation or filtration, and extracted with organic solvents (Lang, 1999; Mimee et al., 2009b).

The effective method of further purification is reprecipitation of glycolipids with distilled water after evaporation of the primary extracts. This technique in some cases yields almost pure cellobiose lipids (Kulakovskaya et al., 2010).

Then, if necessary, thin-layer or column chromatography is used, including high-performance liquid chromatography (HPLC), with organic solvents, such as methanol, chloroform, or their mixtures.

Some examples of glycolipid purification methods illustrating these general principles are given below. The purification methods are presented in more detail in the Appendix.

Cellobiose lipids were obtained as follows (Golubev et al., 2001; Puchkov et al., 2001, 2002): after biomass separation, the culture liquid was lyophilized and the lyophilisate was extracted with methanol. The methanol extract was filtered, evaporated under vacuum, and suspended in water; the water-insoluble component was separated by filtration and redissolved in methanol. Further purification was performed by thin-layer chromatography on Silica gel plates (see the Appendix).

Flocculosin was purified extraction from lyophilized culture liquid with 80% methanol and separation of undissolved components by filtration, followed by evaporation in a rotor evaporator (Cheng et al., 2003). The extract was fractionated using reversed-phase chromatography and elution at different water/methanol ratios. Thin-layer chromatography was used for further purification. However, a simpler method of flocculosin production was developed (Mimee et al., 2009b).

Glycolipids were also extracted from culture liquid using ethyl acetate at a ratio of 1:1. This method was used to obtain MEL (Hewald et al., 2006; Morita et al., 2011b,c), sophorolipids (Lang, 1999; Van Bogaert et al., 2007a), and cellobiose lipids (Morita et al., 2011b,c). In the latter case, purification was performed in a Silica gel column using the chloroform/acetone gradient (10:0 to 0:10) for elution (Morita et al., 2011b,c).

Sophorolipids were extracted by two methods (Cutler and Light, 1979). The minor quantities used for glycolipid production monitoring were extracted as follows: a small amount of cell suspension was separated from the cell culture and extracted with the chloroform/methanol mixture (1:1) for 12 h. The solution was filtered through filter paper, followed by the addition of 20 ml of chloroform and 10 ml of 0.1 N sulfuric acid or distilled water acidified with acetic acid. After separation, the organic layer was harvested and evaporated under the nitrogen. Large amounts of these compounds were extracted by separating the biomass with the cells and the glycolipid precipitate by centrifugation and removal of the culture medium. The cells were suspended in the chloroform/methanol mixture (2:1) and held overnight. The cells were separated by filtration and the organic phase was usually rinsed with water acidified with acetic acid to remove water-soluble components. Organic solvents were removed by nitrogen flow or under vacuum. Further purification was performed by TLC (Thin-Layer Chromatography) or HPLC (Daniel et al., 1999; Otto et al., 1999).

2.3 THIN-LAYER CHROMATOGRAPHY SYSTEMS FOR GLYCOLIPID DETECTION

Thin-layer chromatography of glycolipids is performed on Silica gel plates, mainly from Merck (Germany). The application of Kieselgel $60F_{254}$ plates of the same company with preapplied fluorochrome makes it possible to determine the location of stains under UV illumination. We have used this method to obtain purified preparations; the stains were scraped off and eluted from Silica gel with methanol. Table 2.2 shows some of the solvent systems used for the thin-layer chromatography of glycolipids.

The chromatograms of glycolipids are stained for analytical purposes by moistening or spraying with 5% sulfuric acid solution in ethanol and heating to $\sim 200°C$. The following detection reagent is also used: 10.5 ml α-naphthol (15% in ethanol), 6.5 ml sulfuric acid, 40.5 ml ethanol, and 4 ml of distilled water. MEL was detected on the chromatograms using 0.2% anthrone in 75% sulfuric acid; the chromatogram was sprayed with the solution and heated to 110°C (Lang, 1999).

2.4 CHEMICAL METHODS

It is not our task to analyze the chemical methods of glycolipid study. Carbohydrate and fatty acid analyses are performed mainly for the primary characterization of preparations or in case of considerable diversity of fatty acid residues. Chemical methods of structure analysis, including deacetylation, acid methanolysis, esterification, and gas—liquid chromatography are well described in old papers (Lemieux et al., 1951; Gorin et al., 1961; Fluharty and O'Brien, 1969; Bhattacharjee et al., 1970). For sugar analysis, the samples are hydrolyzed in CF_3CO_2H, and for

Table 2.2 Solvent Systems for the Thin-Layer Chromatography of Yeast Extracellular Glycolipids

Glycolipid	Solvent System (v:v:v)	References
Cellobiose lipids	Chloroform:methanol:water (4:4:0.2) Chloroform:methanol:water (5:3:0.2)	Kulakovskaya et al. (2004, 2009)
Cellobiose lipids	Chloroform:methanol:water (75:25:2)	Morita et al. (2011a)
Sophorolipids	Chloroform:methanol:water (65:15:2)	Lang (1999)
Sophorolipids	Chloroform:methanol (8:2)	Konishi et al. (2008)
MEL	Chloroform:methanol:water (65:15:2)	Lang (1999)
MEL	Chloroform:methanol:7N ammonia (65:15:2)	Morita et al. (2008a–d, 2011b)

fatty acid analysis, they are hydrolyzed by methanolysis. Some methods of chemical modification and the obtaining of full synthetic cellobiose lipid analog are well described (Kulakovskaya et al., 2009).

The enzymatic methods are used for glycolipid modification. For example, enzymatic conversion of diacetylated sophorolipid into acetylated glucoselipid was performed (Imura et al., 2010). The methyl esters, after reacetylation with vinyl acetate using an immobilized lipase, were transesterified with 1,2-3,4-di-O-isopropylidene-D-galactopyranose in tetrahydrofuran using the same lipase catalyst and then the di-O-isopropylidene sophorolipid sugar esters were hydrolyzed to give the galactopyranose sophorolipid esters as the final products (Nunez et al., 2003).

2.5 NMR SPECTROSCOPY AND MASS SPECTROMETRY

Mass spectrometry is a necessary method for glycolipid structure elucidation. Mass spectrometry is performed both by the method of positive or negative ion electrospray (ESI-MS) using the samples dissolved in methanol or pyridine and by the MALDI-TOF/MS (Matrix Assisted Laser Desorption/Ionization) method. Several examples of mass spectra of cellobiose lipid preparations are shown in Figure 2.1. MALDI-TOF MS spectra of sophorolipids and their fatty acids are presented in Konishi et al. (2008).

An ultra-fast liquid chromatography (HPLC) combined with mass spectrometry detection is used for the identification and quantification of glycolipids and their analogs (Hu and Ju, 2001; Ratsep and Shah, 2009).

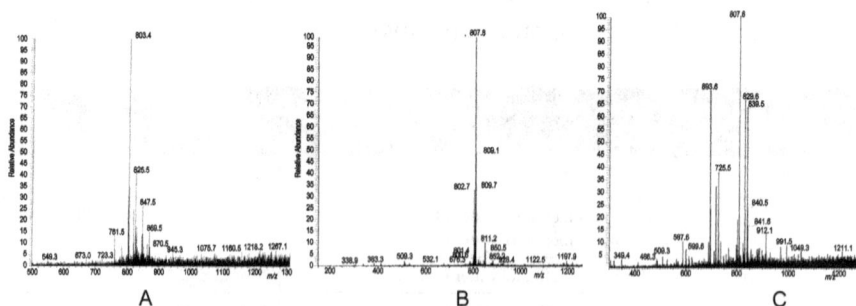

Figure 2.1 Positive ion ESI-MS of cellobiose lipid preparation of (A) Cryptococcus humicola 9-6 *(the major* m/z *signal at 803.4 corresponds to a molecular mass 780.5 plus 22.9 Da for sodium); (B)* Pseudozyma graminicola *VKM Y-2938 (the major* m/z *signal at 807.8 corresponds to a molecular mass 784.9 plus 22.9 Da for sodium); and (C)* Pseudozyma fusiformata *VKM Y-2821 (major* m/z *signals at 807.6 and 693.6 correspond to a molecular masses 784.5 and 670.7 Da, respectively, plus 22.9 Da for sodium).*

Nuclear magnetic resonance (NMR) spectroscopy is an effective method for glycolipid structure elucidation. NMR spectra are recorded in one-dimensional (^{1}H NMR, ^{13}C NMR) and two-dimensional experiments: ^{1}H,^{1}H correlated spectroscopy (COSY); total correlation spectroscopy (TOCSY); rotation frame overhauser effect spectroscopy (ROESY); ^{1}H,^{13}C heteronuclear single quantum coherence (HSQC); and heteronuclear multiple-bond correlation (HMBC). CD_2H or CD_3 (δ_H 3.25 and δ_C 49.0) and tetramethylsilane (TMS) were used as internal standards during signal registration in methanol and pyridine, respectively.

The description of the cellobiose lipid of *Ps. fusiformata* VKM Y-2821 NMR spectroscopic data is given in Table 2.3 as an example. The ^{13}C NMR spectrum contained two signals for the anomeric carbon atoms of sugar residues (δ_C 105.0 and 102.2), three signals of the CO groups (δ_C 179.6, 172.3, and 171.25), signals for CH_3CO (δ_C 20.9) and CH_3-C (δ_C 14.4), $C-CH_2-C$ signals of different intensity (δ_C 43.7−19.5), and two signals of the $O-CH_2-C$ groups

Table 2.3 125-MHz ^{13}C NMR and 500-MHz ^{1}H NMR Chemical Shifts of the Cellobiose Glycolipid of *Pseudozyma fusiformata* VKM Y-2821 (Solution in Pyridine-d_5, Internal TMS as Reference) (Kulakovskaya et al., 2005)

Carbon Atom			Carbon Atom		
	δ_C	δ_H[a]		δ_C	δ_H[a]
C-1	179.6		C-5'	73.15	4.00
C-2	72.3	4.67	C-6'	64.2	4.85; 4.64
C-3	35.9	2.21; 2.09	C-1″	102.2	5.21
C-4	26.3	1.79	C-2″	75.4	5.66
C-5	30.2	1.38	C-3″	76.3	4.26
C-(6-11)	30.15	1.18−1.24	C-4″	71.9	4.12
C-12	30.2	1.28	C-5″	78.8	4.09
C-13	26.3	1.65; 1.49	C-6″	62.5	4.60; 4.21
C-14	34.4	1.64	C-1‴	172.3	
C-15	70.8	4.15	C-2‴	43.7	3.01; 2.99
C-16	75.9	4.20; 3.88	C-3‴	68.2	4.54
C-1'	105.0	4.89	C-4‴	40.3	1.72
C-2'	74.7	4.07	C-5‴	19.5	1.69; 1.55
C-3'	76.3	4.26	C-6‴	14.4	0.90
C-4'	81.95	4.07	CH_3CO	20.7;171.25	2.03
[a]Chemical shifts for the corresponding attached proton(s).					

(δ_C 64.2 and 62.5). Other signals for the O–CH–C groups were located in the region of 68.3–81.95 ppm. The ^{1}H NMR spectrum contained *inter alia* the two doublets characteristic of sugar anomeric protons (δ_H 5.21 and 4.89, $^3J_{1,2}$ 8 Hz), AB spin system of a CH$_2$CO group (δ_H 3.01 and 2.99), protons of a –CH$_2$–CH$_3$ group (triplet at δ_H 0.90, ^{3}J 6 Hz), and a CH$_3$CO group (singlet at δ_H 2.07). The spectrum was assigned using 2D COSY and TOCSY experiments. Analysis of the 2D spectra revealed two residues of β-glucopyranoses, a residue of 2,15,16-trihydroxy-palmitic acid containing 16 carbon atoms, and a residue of 3-hydroxycaproic acid containing six carbon atoms. The 2D ROESY spectrum showed spatial contact of the anomeric proton at δ_H5.21 with the proton H-4 of the other β-glucopyranose residue (δ_H4.07), demonstrating β,1 → 4 linkage between the two residues. The second anomeric proton at δ_H 4.89 proved to be close to H-16 of the 2,15,16-trihydroxy-palmitic acid (correlation peaks δ_H/δ_H 4.07/4.20 and 4.07/3.88). Thus, the cellobiose residue was bound to C-16 of the 2,15,16-trihydroxy-palmitic acid by glycosidic linkage. The assignment in the ^{13}C NMR spectrum was based on the analysis of ^{1}H,^{13}C HSQC and HMBC experiments. The assignment in the HSQC spectrum confirmed the substitution of C-4′ and C-16 due to low-field position of the corresponding signals (δ_C 81.95 and 75.9). The HMBC spectrum contained *inter alia* the following inter- and intra-residue correlation peaks: H-1″/C-4′ (δ_H/δ_C 5.21/81.95), H-1′/C-16 (δ_H/δ_C 4.89/75.9); CH$_3$CO/CH$_3$CO (δ_H/δ_C 2.03/171.25) and H-6′/CH$_3$CO (δ_H/δ_C 4.85/171.25 and 4.64/171.25); H-2″/C-1‴ (δ_H/δ_C 5.66/172.3) and H-2‴/C-1‴ (δ_H/δ_C 3.01/172.3 and δ_H/δ_C 2.99/172.3). The former two correlation peaks confirmed the sequence of the β-glucopyranose residues and the residue of 2,15,16-trihydroxy-palmitic acid as well as positions of the substitution in the residues. Other correlation peaks revealed an *O*-acetyl group at position 6′ of the inner β-glucopyranose residue and 3-hydroxycaproic acid residue at C-2″ of the terminal β-glucopyranose residue. The relatively low-field positions of H-2″ (δ_H 5.66) and H-6′ (δ_H 4.85; 4.64) were in agreement with the well-known effects of *O*-acylation in the ^{1}H NMR spectra of carbohydrates. The ^{1}H and ^{13}C NMR data led to the formula shown in Figure 1.1B.

The NMR data for MELs and sophorolipid of *St. bombicola* are given in the papers of Morita et al. (2007) and Konishi et al. (2008), respectively.

2.6 METHODS FOR STUDYING PHYSICOCHEMICAL PROPERTIES AND ANTIFUNGAL AND MEMBRANE-DAMAGING ACTIVITIES

The methods for determination of various physicochemical properties such as surface tension and the critical concentration of micelle formation have been described in a number of articles (Kitamoto et al., 2002; Puchkov et al., 2002; Konishi et al., 2008; Morita et al., 2008a−d). The adsorption of sophorolipid and their mixtures with the anionic surfactant sodium dodecyl benzene sulfonate has been measured at the air/water interface by neutron reflectivity (Chen et al., 2011). The methods of antifungal activity assay are based on the detection of growth inhibition zones or cell survival. The methods for determining the membrane-damaging activity are described in detail in the Appendix.

2.7 MOLECULAR BIOLOGY METHODS

The data on the biosynthetic pathways of yeast extracellular glycolipids have been obtained by the modern methods of gene identification and cloning, proteomic analysis, and analysis of intermediate compounds formed under mutation in the genes encoding the enzymes of biosynthetic pathways. Not being experts in these methods, we will not dwell in detail on the methodical aspect of these works but give a brief description of the approaches used.

The gene disruption and analysis of glycolipid production in mutant strain is a widely-used approach in the study of biosynthesis of these compounds. To generate mutants of *U. maydis* that are unable to produce MELs, the available genome database of this fungus was searched for putative glycosyltransferases that could be involved in the generation of the central mannosyl-β-δ-*erythritol moiety* (Hewald et al., 2005). About 40 genes encoding proteins with some similarity to glycosyltransferases were identified. The function of some of these enzymes could be derived by similarity to known glycosyltransferases involved in cell wall biosynthesis or protein glycosylation. For the remaining candidate genes, mutants were systematically generated by a PCR-based deletion strategy. After another round of PCR amplification, the replacement constructs were transformed into protoplasts of the haploid *U. maydis* strains. Transformants were checked for successful deletion of the respective genes by Southern analysis. The deletion mutants were shifted to nitrogen starvation, glycolipids were extracted and analyzed by layer chromatography. One of the mutants showed a total loss of MEL production as detected by TLC analysis. The deletion of this putative glycosyltransferase gene *Emt1* completely blocked MEL production (Hewald et al., 2005). In addition, the production of the cellobiose lipid was suppressed by deleting the P450 monooxygenase Cyp1 gene (Hewald et al., 2005).

To identify additional genes of MEL biosynthesis, a genome-wide expression analysis using DNA microarray was performed (Hewald et al., 2006). Under nitrogen limitation which induces MEL production, the induced genes were revealed and glycolipid production in deletion mutant analyzed (Hewald et al., 2006).

To investigate the biosynthesis of MELs in the yeast *Ps. antarctica*, the reported expressed sequence tag analysis and estimating of genes expressing under MEL production were performed (Morita et al., 2010a). Among the genes, a *PaEMT1* gene was revealed with high sequence identity to the gene *emt1*, encoding an erythritol/mannose transferase of *U. maydis*. The obtained Δ*PaEMT1* strain failed to produce MELs, while its growth was the same as that of the parental strain.

The *Uhd1* gene in *U. maydis* was disrupted and forming of cellobiose lipid lacking δ-hydroxyl group of the short-chain fatty acid was observed in the mutant strain (Teichmann et al., 2011a). The biosynthesis gene cluster of *Ps. flocculosa* was revealed by searching of genes homologous to genes of biosynthesis gene cluster of *U. maydis* responsible for ustilagic acid biosynthesis (Teichmann et al., 2011b).

The strategy of identification of *at*, *Ugta1*, and *Ugtb1* genes encoding the enzymes of sophorolipid biosynthesis included the blast-homology search of the genes with high homology to known carbohydrate transacetylases and glycosetransferases and deletion mutants (Saerens et al., 2011a−c). The pure unacetylated sophorolipid produced by mutant strain Δ*at* confirmed the role of *at* product in glycolipid acetylation (Saerens et al., 2011c). The Δ*Ugta1* mutant was unable to produce sophorolipid while Δ*Ugtb1* mutant produces 17-*O*-(β-D-glucopyranosyl)-octadecenoic acid, indicating the roles of both enzymes in fatty acid glycosylation (Saerens et al., 2011a−c).

The complete genomes of *U. maydis* (Kämper et al., 2006), *Ps. antarctica* (Morita et al., 2013), and *St. bombicola* (Van Bogaert et al., 2013) were sequenced and annotation. The availability of these genome sequences gave new keys to studying fungal glycolipid biosynthesis.

CHAPTER 3

Physicochemical Properties of Yeast Extracellular Glycolipids

3.1 SOLUBILITY

According to our observations, cellobiose lipids dissolve in 0.04 M citrate−phosphate buffer, at least to a concentration of 2 g/l. Cellobiose lipids do not dissolve in deionized water. This property was used in the purification process to separate the precipitate of antifungal substances from methanol-soluble salts, sugars, and other compounds present in the culture liquid. They do not dissolve in chloroform but are soluble in pyridine and methanol, at least, to a concentration of 20−50 g/l.

MELs and sophorolipids are soluble in chloroform and ethyl acetate (Lang, 1999); this property is often used for extraction from the culture liquid and purification. They are soluble in methanol and chloroform−methanol mixtures.

Sophorolipids dissolve well in methanol, ethanol, acetonitrile, and dimethyl sulfoxide (DMSO); they disperse in mineral oil, vegetable oil, glycerol, and, propylene glycol, and at pH 5.0 and lower, they disperse in water (Van Bogaert et al., 2011). Crude preparation of sophorolipids was more water soluble (2−3 g/l) when compared to purified preparation (0.07 g/l) (Otto et al., 1999).

3.2 STABILITY DURING STORAGE AND THERMAL STABILITY

The preparations of cellobiose lipids of *Cr. humicola*, *Ps. fusiformata*, and *Ps. graminicola* were stored in our experiments as methanol solutions at 0−5°C for 1.5−2 years without loss of antifungal activities. The cellobiose lipids of *Cr. humicola* were shown to maintain the activity under heating to 50°C for at least 2−3 h and to 100°C for 30 min (Golubev and Shabalin, 1994).

Sophorolipids preserve their surface-active properties at high salt concentration (Hirata et al., 2009) and in a wide temperature range (Nguyen et al., 2010). Under long-term storage at pH values higher than 7.0–7.5, the irreversible hydrolysis of the acetyl groups and ester bonds was observed (Van Bogaert et al., 2011).

3.3 MOLECULAR MASSES

The molecular masses of extracellular yeast glycolipids vary due to different O-substituents in the sugar residue and the number of hydroxyl groups and carbon atoms in fatty acid residues. The molecular masses of some cellobiose lipids and sophorolipids are given in Tables 3.1 and 3.2, respectively. The molecular masses of MELs are not presented, because these compounds are even more variable. The structural variants of MEL from various yeast species are presented in review (Arutchelvi et al., 2008).

3.4 SURFACE-ACTIVE PROPERTIES

Surface tension and critical micelle concentration (CMC) are physical values that characterize surface-active properties of compounds. Due to the application of natural glycolipid mixtures in many cases, these

Table 3.1 Molecular Masses of Cellobiose Lipids of Some Yeast		
Species and Strains	Formulae	Molecular Mass
Cryptococcus humicola 9-6, VKM Y-1613, VKM Y-2238	$C_{36}H_{60}O_{18}$	780
Cryptococcus humicola X—397 Trichosporon porosum VKM Y-2956	$C_{36}H_{60}O_{18}$ and $C_{34}H_{58}O_{17}$ $C_{36}H_{60}O_{18}$ and $C_{34}H_{58}O_{17}$	780 and 738
Cryptococcus humicola X—297	$C_{34}H_{58}O_{17}$	738
Pseudozyma fusiformata VKM Y-2821, VKM Y-2898, Ll-41, PTZ-356	$C_{30}H_{54}O_{16}$ and $C_{36}H_{64}O_{18}$	670 and 784
Pseudozyma fusiformata VKM Y-2909, Ll-71, PTZ-351 Pseudozyma graminicola Ll-46, VKM Y-2938	$C_{36}H_{64}O_{18}$	784
Pseudozyma fusiformata Ll-16	$C_{36}H_{64}O_{18}$ and $C_{38}H_{66}O_{19}$	784 and 826
Sympodiomycopsis paphiopedili VKM Y-2817	$C_{28}H_{52}O_{15}$	628
Source: Kulakovskaya et al. (2005, 2006, 2010) and Golubev et al. (2008a,b).		

Table 3.2 Molecular Masses of Some Sophorolipids (SL)		
Structural Features	Formula	Molecular Mass
Nonacetylated lactonic SL with C18:1 fatty acid	$C_{30}H_{52}O_{12}$	604
Nonacetylated acidic SL with C18:2 fatty acid	$C_{30}H_{52}O_{13}$	620
Nonacetylated acidic SL with C18:1 fatty acid	$C_{30}H_{54}O_{13}$	622
Nonacetylated acidic SL with C18:0 fatty acid	$C_{30}H_{56}O_{13}$	624
Monoacetylated acidic SL with C16:1 fatty acid	$C_{30}H_{52}O_{14}$	636
Monoacetylated acidic SL with C16:0 fatty acid	$C_{30}H_{54}O_{14}$	638
Monoacetylated lactonic SL with C18:2 fatty acid	$C_{32}H_{52}O_{13}$	644
Monoacetylated lactonic SL with C18:1 fatty acid	$C_{32}H_{54}O_{13}$	646
Monoacetylated acidic SL with C18:2 fatty acid	$C_{32}H_{54}O_{14}$	662
Monoacetylated acidic SL with C18:1 fatty acid	$C_{32}H_{56}O_{14}$	664
Monoacetylated acidic SL with C18:0 fatty acid	$C_{32}H_{58}O_{14}$	666
Diacetylated acidic SL with C16:2 fatty acid	$C_{32}H_{52}O_{15}$	676
Diacetylated acidic SL with C16:1 fatty acid	$C_{32}H_{54}O_{15}$	678
Diacetylated acidic SL with C16:0 fatty acid	$C_{32}H_{56}O_{15}$	680
Diacetylated lactonic SL with C18:2 fatty acid	$C_{34}H_{54}O_{14}$	686
Diacetylated lactonic SL with C18:1 fatty acid	$C_{34}H_{56}O_{14}$	688
Diacetylated acidic SL with C18:2 fatty acid	$C_{34}H_{56}O_{15}$	704
Diacetylated acidic SL with C18:1 fatty acid	$C_{34}H_{58}O_{15}$	706
Source: Kurtzman et al. (2010) and Ma et al. (2011).		

values are expressed not only as molar concentrations but also as weight concentrations. Some data are given in Tables 3.3 and 3.4.

The surface-active properties of sophorolipids are described in detail in Zhang et al. (2004) and Hirata et al. (2009). Sophorolipid methyl, ethyl, propyl, and butyl esters were obtained in the work (Zhang et al., 2004), and esterification was shown to reduce the CMC. It is significant for application of amphiphilic compounds as detergents.

The nonacetylated lactonic sophorolipid shows increased foam formation and water solubility when compared to the acetylated form (Saerens et al., 2011b).

Purified sophorolipids were more surface active (CMC, 10 mg/l) than with crude preparation (CMC, 130 mg/l).

The structure of sugar moiety of sophorolipids shows little effect on surface properties. In the study of the diacetylated sophorolipid

Table 3.3 The Surface Tension of Glycolipids and Some Known Detergents

Compound	Surface Tension (mN/m)	Assay Conditions	References
Sophorolipids, acidic form	30	H_2O, 25–40°C	Lang (1999)
Sophorolipids, lactonic form	35	H_2O, 25–40°C	Lang (1999)
Sophorolipids mixture	37.2		Hirata et al. (2009)
Sophorolipid of *Candida batistae*	39.3	H_2O	Konishi et al. (2008)
Sophorolipid of *Starmerella bombicola*	43.2	H_2O	Konishi et al. (2008)
MEL	28	H_2O, 25–40°C	Lang (1999)
MEL-A	28.4		Kitamoto et al. (1993)
MEL-B	28.2	25°C	Kitamoto et al. (1993)
MEL-B	25.2	25°C	Morita et al. (2011b)
MEL-C	30.7		Morita et al. (2008b)
Mannosyl mannitol lipid	24.2		Morita et al. (2009a,b)
Cellobiose lipid of *Cryptococcus humicola*	37	0.1 M $NaHCO_3$ 23°C	Puchkov et al. (2002)
Rhamnolipid	25	H_2O, 25–40°C	Lang (1999)
Sodium dodecyl sulfate (SDS)	32	0.1 M $NaHCO_3$ 23°C	Puchkov et al. (2002)
DTAB (Dodecyltrimethylammonium bromide)	51	0.1 M $NaHCO_3$ 23°C	Puchkov et al. (2002)

Table 3.4 Critical Micelle Concentration of Some Glycolipids

Compound	CMC	References
Sophorolipid diacetylated C18, lactonic form	366 mg/l	Konishi et al. (2008)
Sophorolipid diacetylated C18, acidic form	138 mg/l	Konishi et al. (2008)
Sophorolipid nonacetylated C18, lactonic form	95 mg/l	Konishi et al. (2008)
Sophorolipid nonacetylated C18, acidic form	17 mg/l	Konishi et al. (2008)
Sophorolipid mixture	43 mg/l	Hirata et al. (2009)
MEL-A	2.7×10^{-6} M	Kitamoto et al. (1993)
MEL-B	4.5×10^{-6} M	Kitamoto et al. (1993)
MEL-B	3.7×10^{-6} M	Morita et al. (2011b)
MEL-C	4.5×10^{-6} M	Morita et al. (2008b)
Mannosyl mannitol lipid	2.6×10^{-6} M	Morita et al. (2009a,b)
Cellobiose lipid of *Cryptococcus humicola*	2×10^{-5} M (pH 4.0)	Puchkov et al. (2002)
Cellobiose lipid of *Cryptococcus humicola*	3.3×10^{-5} M (pH 4.0)	Morita et al. (2011a)
	4.1×10^{-4} M (pH 7.0)	

17-[(2′-O-β-glycopyranosyl-β-D-glycopyranosyl)oxy]-*cis*-9-oktadecenoate 6′6″-diacetate, its nonacetylated derivative and enzymatically obtained 17-[(β-D-glucopyranosyl)oxy]-*cis*-9-octadecenoate 6′-acetate and 17-[(β-D-glycopyranosyl)oxy]-*cis*-9-oktadecenoate showed similarity of interfacial parameters: the CMC was $1.7 - 1.6 \times 10^{-4}$ M and surface tension was nearly 40 mN/m (Imura et al., 2010).

3.5 LACTONIZATION AND SELF-ASSEMBLY

One of the properties of sophorolipids is the ability for nonenzymatic lactonization. Sophorolipids can occur in the acidic form with free carboxyl group or in the lactonic form with the internal etherification between carboxylic group of fatty acid residue and 4″−OH group of sophorose residue. Lactonic sophorolipids have better surface tension lowering and antimicrobial activity, while the acidic forms possess a better foam production and solubility (Van Bogaert et al., 2011).

Lactonization of many sophorolipid molecules at acidic pH values may result in formation of giant helical chains of 5−11 μm in width and up to several hundreds of micrometers in length (Zhou et al., 2004). Neutralization of pH values slows down the formation of "bands" and leads to the formation of shorter aggregates.

The self-assembly of sophorolipid of *St. bombicola* was investigated using a combination of small-angle neutron scattering (SANS), transmission electron microscopy under cryogenic conditions (Cryo-TEM), and NMR techniques, and found a strong dependence of glycolipid self-assembly on the degree of ionization of the COOH group at concentration values as low as 5 and 0.5 wt% (Baccile et al., 2012). At a low degree of ionization, self-assembly was driven by concentration, and micelles were mainly nonionic; at mid degree of ionization, the formation of COO(−) groups introduces negative charges at the micellar surface; at full ionization, large netlike tubular aggregates appeared (Baccile et al., 2012).

MELs in aqueous suspensions can also exist as micelles or other self-organized structures. MEL can form giant vesicles of more than 10 μm in diameter (Kitamoto et al., 2000, 2002; Imura et al., 2006). The ability of MELs to form vesicles of different sizes and structures, coacervates and monolayers at the surfaces of various solvents is described in detail in the reviews (Kitamoto et al., 2002; Arutshelvi et al., 2008). Such properties have also been studied for mannosyl mannitol lipid (Morita et al., 2009b). The diastereomers of mannosylerythritol lipids differ in interfacial properties and aqueous phase behavior (Fukuoka et al., 2012).

3.6 INTERACTION BETWEEN CELLOBIOSE LIPIDS AND ARTIFICIAL MEMBRANES

The interaction between cellobiose lipids of *Cr. humicola* and artificial membranes was studied (Puchkov et al., 2002). The fluorescence resonance energy transfer (FRET) method with specific fluorescent probes demonstrated that the fluorescent signal of liposomes obtained from the diphthanoyl phosphatidylcholine:phosphatidylserine:ergostrol mixture (20:2:0.5) changes with the addition of cellobiose lipids in the same manner as on addition of lysophosphatidylcholine, known for its ability to be incorporated into liposomes. It was also shown that this preparation could cause fluctuations in electrical conductivity of artificial lipid bilayers of different compositions and the breakage of these bilayers at higher concentrations. The cellobiose lipid differed in the character of induced fluctuations from the known membranedamaging agent nystatin (Ng et al., 2003) and from the potassium carrier nonactin, which could be indicative of differences in some details of the mechanism of action. These experiments suggested that cellobiose lipids are incorporated into lipid bilayers and, at low concentrations, cause the formation of short-living channels of different sizes. The higher cellobiose lipid concentrations induce more considerable damages (Puchkov et al., 2001, 2002).

Thus, yeast extracellular glycolipids are surface-active amphiphilic compounds capable of self-organization of supramolecular structures in different solvents and interaction with lipid bilayers.

CHAPTER 4

Biological Activities of Extracellular Yeast Glycolipids

4.1 ANTIFUNGAL ACTIVITY OF CELLOBIOSE LIPIDS

4.1.1 Discovery of Antifungal Activity of Cellobiose Lipids

The ability of cellobiose lipids to inhibit the growth of microorganisms, including bacteria and fungi, was first revealed in *U. maydis* (Haskins and Thorn, 1951). However, the antibiotic activity was low.

The new interest in the study of antifungal properties of extracellular glycolipids emerged in 1994, when *Cr. humicola* was shown to have a low-molecular protease-insensitive antifungal agent secreted into the liquid medium and inhibiting the growth of various species of yeast and fungi (Golubev and Shabalin, 1994). The antifungal activity manifested only at acidic pH. The "culture-to-culture" method of primary detection of antifungal activity (see the Appendix A.2.1) allows the screening of promising producers. Figure 4.1 shows the results of such testing. By this approach it was shown that *Cr. humicola* (Golubev and Shabalin, 1994), *Ps. fusiformata* (Golubev et al., 2001), *Symp. paphio-pedili* (Golubev et al., 2004), *Ps. graminicola* (Golubev et al., 2008b), and *Tr. porosum* (Golubev et al., 2008a) suppress the growth of yeast and fungi belonging to different systematic groups. Later it was demonstrated that cellobiose lipids are responsible for the antifungal activities of the above yeast species (Puchkov et al., 2002; Kulakovskaya et al., 2004, 2005, 2010; Golubev et al., 2008a,b) and of *Ps. flocculosa* (Mimee et al., 2005).

4.1.2 The Spectrum of Cellobiose Lipid Antimicrobial Activity

The yeasts secreting cellobiose lipids inhibit the growth of a wide range of both ascomycetes and basidiomycetes, including medically important species of the genera *Candida*, *Filobasidiella*, *Malassezia*, *Trichosporon* and phytopathogenic species of the genera *Diaporthe*, *Entyloma*, *Farysia*, *Sclerotinia*, *Phomopsis*, and *Ustilago*. Cellobiose lipids have a much broader activity spectrum compared to killer toxins, which are characterized by high specificity and inhibit the growth

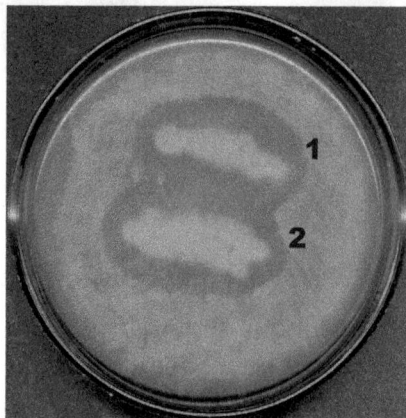

Figure 4.1 The "culture-to-culture" testing of antifungal activity. Culture medium contains (g/l): glucose, 5.0; peptone, 2.5; yeast extract, 2; agar, 20; citric acid × H₂O, 6.5; Na₂HPO₄ × 12H₂O, 13.8 (pH 4.0). The medium was preinoculated with Cryptococcus terreus and then Cryptococcus humicola 9-6 (1) and Trichosporon porosum VKM Y-2956 (2) were streaked. Petri dishes were incubated at 24°C for 3 days.

of closely related species (Golubev, 2006). Table 4.1 lists the spectra of antifungal action of the yeast *Cr. humicola* (Golubev and Shabalin, 1994), *Ps. fusiformata* (Golubev et al., 2001), *Symp. paphiopedili* (Golubev et al., 2004), and *Ps. graminicola* (Golubev et al., 2008b) obtained by "culture-to-culture" method.

Some of the cultures insensitive to cellobiose lipids according to "culture-to-culture" method actually demonstrate the sensitivity when using the chromatographically pure preparations of glycolipids. Insensitivity to these agents revealed by the culture-to-culture testing may be due to insufficient concentration of cellobiose lipids secreted into the medium by the producer under study. To date, the insensitivity of yeasts to cellobiose lipids has been reliably demonstrated only for their producers. It is most likely associated with the peculiarities of their membrane structures and (as will be described in the chapter concerned with metabolism of these compounds) with the ability for degradation of these compounds.

To study the antifungal activity of pure preparations, we used, along with *Saccharomyces cerevisiae* and *Cr. terreus*, the yeast species known as pathogens (Hurley et al., 1987): *Filobasidiella neoformans* and some species of genera *Trichosporon* and *Candida*. Growth inhibition zones on agarized media under the influence of different quantities of purified cellobiose lipid of *Ps. graminicola* are shown in Figure 4.2.

Table 4.1 The Spectra of Antifungal Action of *Cryptococcus humicola* (Golubev and Shabalin, 1994), *Pseudozyma fusiformata* (Golubev et al., 2001), *Pseudozyma graminicola* (Golubev et al., 2008b), and *Sympodiomycopsis paphiopedili* (Golubev et al., 2004) Under Testing by "Culture-to-Culture" Method

Genera	Strains Producing Antifungal Glycolipids			
	Cryptococcus humicola 9-6	*Pseudozyma fusiformata* VKM Y-2821	*Pseudozyma graminicola* VKM Y-2938	*Sympodiomycopsis paphiopedili* VKM Y-2817
Basidiomycetes				
Agaricostilbum	(1,1)	(1,1)		(1,1)
Apiotrichum	(1,1)			
Atractogloea	(1,1)	(0,0)		(0,0)
Bensingtonia	(6,7)	(6,6)	(8,8)	(6,6)
Bullera	(8,9)	(11,30)	(5,5)	(11,28)
Bulleromyces	(1,4)	(1,1)		(1,1)
Calocera	(1,1)	(0,0)		
Christiansenia	(1,1)			
Cryptococcus	(29, 121)	(33,33)		(28,37)
Cystofilobasidium	(1,4)	(4,4)		(3,3)
Dioszegia			(2,2)	(2,9)
Endophyllum	(1,1)	(0,0)		(1,1)
Entyloma	(1,1)	(1,1)	(1,1)	(0,0)
Erythrobasidium		(1,1)	(1,1)	(1,1)
Exobasidium	(4,4)	(6,8)	(6,8)	(4,5)
Farysia	(0,0)	(1,1)	(1,1)	(1,1)
Fellomyces	(1,1)	(3,3)	(1,1)	(3,3)
Fibulobasidium	(1,1)		(1,1)	
Filobasidiella	(1,3)	(1,44)	(1,2)	(1,50)
Filobasidium	(4,8)	(2,4)	(1,1)	(3,3)
Guepiniopsis	(0,0)	(0,0)		(0,0)
Guehomyces			(1,2)	
Gymnosporangium	(1,1)	(1,1)		(1,1)
Holtermannia		(1,1)	(1,1)	(1,1)
Itersonilia	(1,6)	(1,2)	(1,1)	(1,2)
Kondoa	(1,2)		(1,1)	
Kockovaella		(1,1)	(1,1)	(1,1)
Kurtzmanomyces	(1,1)	(1,1)		(1,1)
Leucosporidium	(3,8)	(2,2)	(2,2)	(2,2)
Malassezia		(2,2)		(2,2)

(Continued)

Table 4.1 (Continued)

Genera	Strains Producing Antifungal Glycolipids			
	Cryptococcus humicola 9-6	*Pseudozyma fusiformata* VKM Y-2821	*Pseudozyma graminicola* VKM Y-2938	*Sympodiomycopsis paphiopedili* VKM Y-2817
Mastigobasidium		(1,1)	(1,1)	(1,1)
Microbotryium	(4,4)	(4,4)	(5,7)	(5,7)
Microstroma	(1,1)			
Mrakia	(2,2)	(2,3)	(2,3)	(2,4)
Neovossia	(0,0)	(0,0)		(0,0)
Platygloea	(0,0)	(1,1)		(0,0)
Phaffia	(1,17)			(1,1)
Pseudozyma		(0,0)	(5,11)	(3,8)
Puccinia	(2,2)	(1,1)	(2,2)	(1,1)
Rhodosporidium	(8,19)	(1,1)	(8,15)	(4,4)
Rhodotorula	(16,103)	(34,43)	(16,44)	(37,46)
Sakaguchia			(1,2)	
Sebacina	(0,0)			
Septobasidium	(1,1)	(1,1)		(1,1)
Sirobasidium	(1,2)		(1,1)	
Sorosporium		(1,1)		(1,1)
Sphacelotheca	(1,1)	(1,1)	(1,1)	(1,1)
Sporidiobolus	(2,6)	(3,3)	(5,6)	(4,4)
Sporisorium	(1,1)	(2,3)	(1,1)	(2,3)
Sporobolomyces	(10,19)	(25,25)	(3,3)	(24,26)
Sterigmatomyces	(1,2)	(2,2)	(2,3)	(2,3)
Sterigmatosporidium	(0,0)		(1,3)	
Sympodiomycopsis		(1,1)	(1,1)	
Tausonia		(1,1)		
Tilletia	(1,1)		(1,1)	(1,1)
Tilletiaria		(1,1)		(1,1)
Tilletiopsis	(0,0)	(7,9)	(3,4)	(5,6)
Tremella	(8,8)		(2,2)	
Trichosporon	(6,19)	(11,12)	(9,9)	(12,13)
Tsuchiyaea				(1,1)
Udeinomyces				(3,12)
Ustilago	(4,4)	(10,12)	(6,7)	(7,7)
Xanthophyllomyces		(1,4)	(1,2)	(1,4)

(Continued)

Table 4.1 (Continued)

Genera	Strains Producing Antifungal Glycolipids			
	Cryptococcus humicola 9-6	*Pseudozyma fusiformata* VKM Y-2821	*Pseudozyma graminicola* VKM Y-2938	*Sympodiomycopsis paphiopedili* VKM Y-2817
Ascomycetes				
Aciculoconidium	(0,0)			
Ambrosiozyma	(1,1)		(4,4)	
Arthroascus	(2,4)	(1,1)	(2,2)	(1,1)
Arxiozyma	(0,0)			
Arxula		(1,1)		(1,1)
Brettanomyces	(0,0)			
Candida	(0,0)	(2,2)	(8,9)	(3,3)
Citeromyces	(1,4)	(0,0)	(1,1)	(1,1)
Clavispora	(1,2)	(0,0)	(1,1)	(1,1)
Debaryomyces	(8,10)	(7,7)	(10,12)	(7,9)
Dekkera	(1,1)		(1,1)	(1,1)
Diaporthe				(1,1)
Dipodascus		(1,1)		(1,1)
Endomyces				(1,1)
Geotrichum	(1,1)		(1,1)	
Guiliermondella	(1,1)	(1,1)	(1,1)	(1,1)
Hanseniaspora		(0,0)	(2,2)	
Hormoascus	(1,2)			
Hyphopichia	(1,1)			
Issatchenkia	(2,4)	(1,1)		(1,1)
Kazachstania			(6,6)	
Kloeckera	(1,1)			
Kluyveromyces	(1,1)		(3,5)	(4,5)
Komagataea			(1,1)	
Komagataella			(1,1)	
Lipomyces	(3,11)	(3,3)	(2,2)	(2,2)
Lodderomyces	(1,1)	(1,1)	(1,1)	(1,1)
Mastigomyces	(2,3)	(1,1)	(2,2)	(1,1)
Metschnikowia	(8,29)		(8,22)	(4,4)
Myxozyma	(1,1)			(4,4)
Nadsonia	(2,25)	(1,1)	(2,3)	(1,1)
Nakaseomyces			(1,1)	
Nakazawaea			(1,1)	

(Continued)

Table 4.1 (Continued)

Genera	Strains Producing Antifungal Glycolipids			
	Cryptococcus humicola 9-6	*Pseudozyma fusiformata* VKM Y-2821	*Pseudozyma graminicola* VKM Y-2938	*Sympodiomycopsis paphiopedili* VKM Y-2817
Naumovia			(1,1)	
Nematospora	(1,1)		(1,1)	
Oosporidium	(1,1)			
Pachysolen	(1,1)		(1,1)	
Pachitychospora	(1,1)			
Phomopsis			(1,1)	
Pichia	(55,61)	(1,1)	(4,4)	(1,1)
Protomyces	(0,0)	(1,1)	(2,2)	(2,2)
Saccharomyces	(1,4)	(1,1)	(5,6)	(0,0)
Saccharomycodes	(1,4)		(1,4)	(0,0)
Saccharomycopsis	(1,1)		(1,1)	
Saturnospora	(2,5)	(0,0)	(1,1)	(1,1)
Schizoblastosporion	(1,15)			
Schizosaccharomyces	(2,3)	(0,0)	(2,4)	(2,6)
Schwanniomyces			(1,1)	
Sclerotinia			(1,1)	(1,1)
Sporopachydermia	(1,1)			
Stephanoascus	(1,1)	(1,1)		(1,1)
Taphrina	(13,13)	(8,8)	(7,7)	(6,6)
Tetrapisispora			(1,1)	
Torulaspora	(1,1)	(1,1)	(3,3)	(1,1)
Trigonopsis	(1,2)			
Wickerhamia	(1,1)	(1,1)	(1,1)	(1,1)
Wickerhamiella	(1,1)		(1,1)	
Williopsis	(2,2)	(0,0)	(1,1)	(1,1)
Wingea	(0,0)			
Yamadazyma	(1,1)			
Yarrowia	(1,8)		(1,9)	(0,0)
Zygoascus	(0,0)		(1,2)	(1,1)
Zygosaccharomyces	(1,1)	(0,0)	(1,2)	(1,1)
Zygozyma		(1,1)		(1,1)
Zygotorulaspora			(1,1)	
Zygowilliopsis			(1,1)	

The numbers of sensitive species and strains are indicated in parenthesis.

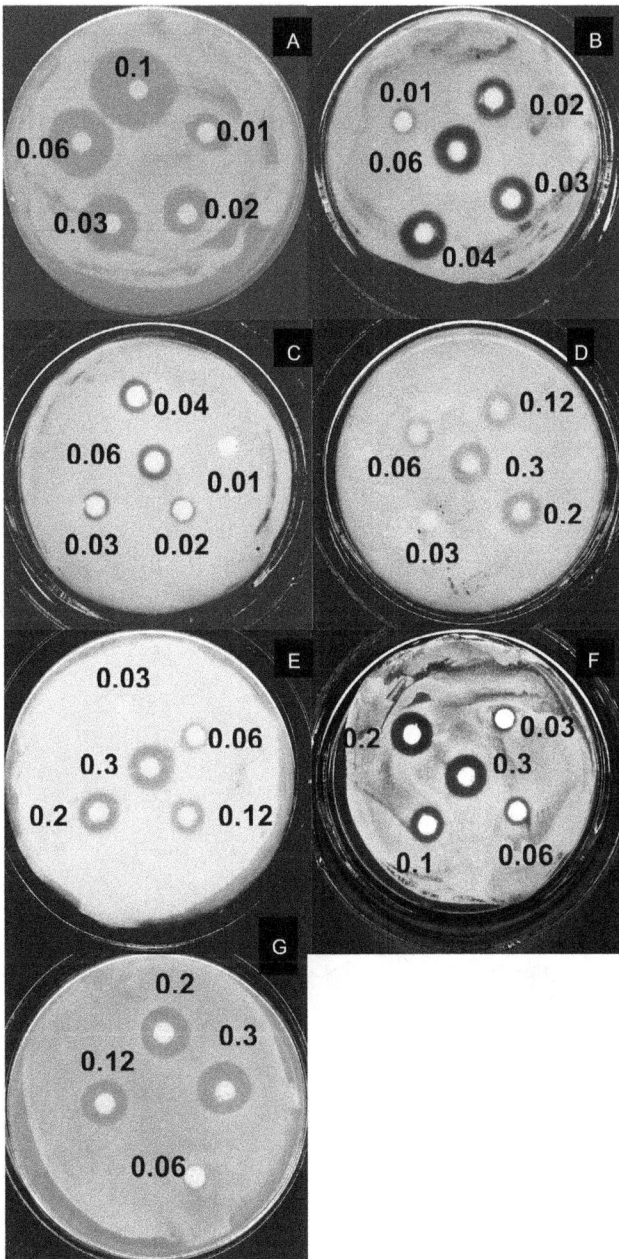

Figure 4.2 The inhibition of yeast growth by purified cellobiose lipid of Pseudozyma graminicola VKM Y-2938. Culture medium contains (g/l): glucose, 5.0; peptone, 2.5; yeast extract, 2; agar, 20; citric acid × H₂O, 6.5; Na₂HPO₄ × 12 H₂O, 13.8 (pH 4.0). Petri dishes were incubated at 24°C for 3 days. Numbers indicate the amounts of cellobiose lipid (mg per disc) placed on Whatman GF glass fiber discs. Test cultures: Cryptococcus terreus VKM Y-2253 (A), Filobasidiella neoformans IGC 3957 (B), Candida viswanathii CBS 4024 (C), Candida glabrata CBS 138 (D), Clavispora lusitaniae IGC 2705 (E), Candida albicans JCM 1542 (F), Saccharomyces cerevisiae VKM Y-1173 (G).

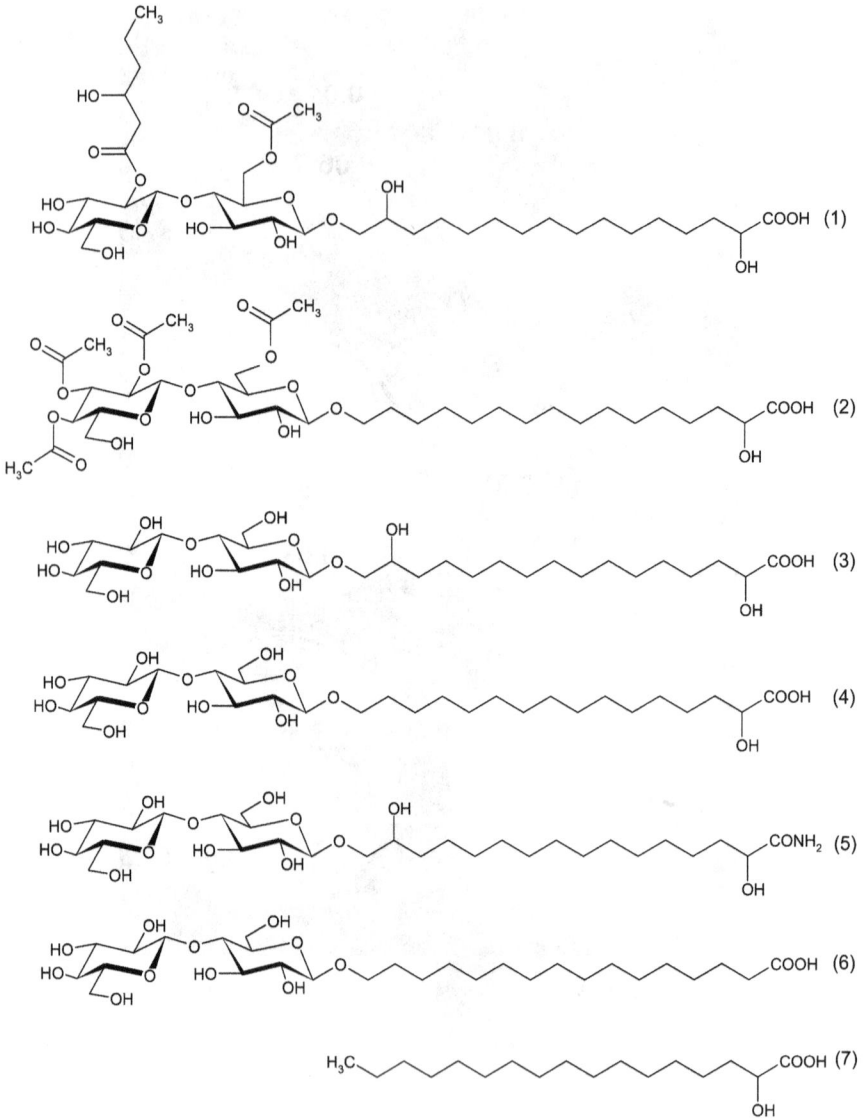

Figure 4.3 The structures of cellobiose lipids of Pseudozyma fusiformata and Pseudozyma graminicola (1), Cryptococcus humicola (2), deacylated derivatives of cellobiose lipids (3 and 4), acetamide of deacylated glyco-lipid 3 (5), synthetic compound 16-(β-cellobiosyloxy)-hexadecanoic acid (6), and 16-hydroxyhexadecanoic acid (7).

We compared the antifungal activities of glycolipids of *Cr. humicola* and *Pseudozyma*, which are different in the structures of O-substituents (Figure 4.3). The antifungal activity assay by the method of growth inhibition on an agarized medium for the test

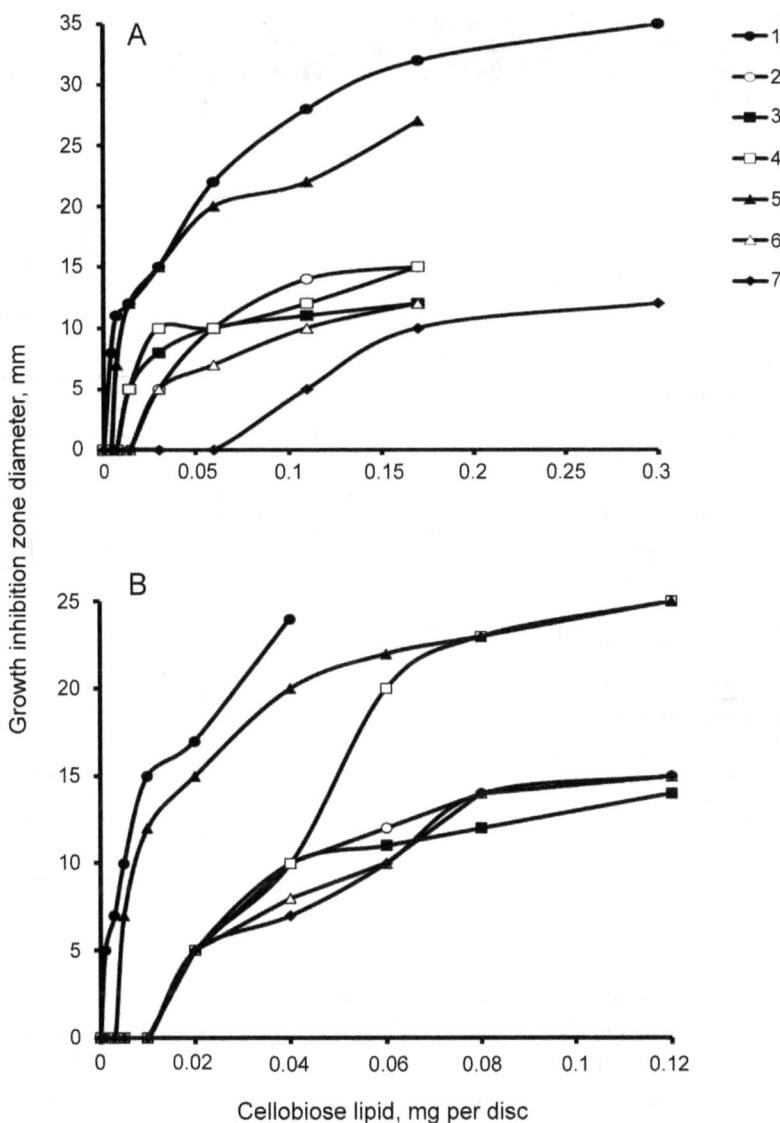

Figure 4.4 The diameters of growth inhibition zones under action of purified cellobiose lipid of Cryptococcus humi-cola 9-6 (A) and Pseudozyma graminicola VKM Y-2938 (B) at pH 4.0. Test cultures: Cryptococcus terreus VKM Y-2253 (1), Clavispora lusitaniae IGC 2705 (2), Candida glabrata CBS 138 (3), Candida viswanathii CBS 4024 (4), Filobasidiella neoformans IGC 3957 (5), Candida albicans JCM 1542 (6), Saccharomyces cere-visiae VKM Y-1173 (7).

cultures under study showed no difference between two cellobiose lipids (Figure 4.4). The producers showed no sensitivity to their own glycolipids, even with 0.5 mg of glycolipids applied to Petri dishes.

Table 4.2 Cell Viability Under Action of Purified Cellobiose Lipid of *Cryptococcus humicola* 9-6

Yeast Strains	Cellobiose Lipid Concentration, mg/ml									
	0	0.006	0.012	0.02	0.05	0.08	0.15	0.20	0.30	0.50
	Cell Viability, % of Control									
Cryptococcus terreus VKM Y-2253	100	–	–	2	0	0	–	–	–	–
Filobasidiella neoformans IGC 3957	100	–	–	5	1	0	–	–	–	–
Trichosporon asteroides VKM Y-275	100	86	67	43	7	3	0	–	–	–
Trichosporon asahii VKM Y-803	100	22	17	15	2	1	0	–	–	–
Trichosporon faecalis VKM Y-2865	100	100	97	81	10	7	0	–	–	–
Candida viswanathii CBS 4024	100	–	–	71	16	10	–	–	–	–
Candida tropicalis RBF-988	100	–	–	–	100	48	27	10	6	0
Candida parapsilosis VKM Y-58	100	–	–	–	100	52	36	5	3	1
Candida glabrata CBS 138	100	–	–	100	–	–	25	–	2	1.5
Candida albicans JCM 1542	100	–	–	100	–	–	2	–	1	0
Clavispora lusitaniae IGC 2705	100	–	–	100	–	–	10	–	1	1
Saccharomyces cerevisiae VKM Y-1173	100	–	–	100	–	–	5	–	1	0

The cells of test cultures were treated in 0.04 M citrate–phosphate buffer, pH 4.0, at room temperature for 30 min.

The survival rates of various species of yeast fungi, including the candidiasis and cryptococcosis pathogens, were determined during the treatment with cellobiose lipids in a liquid medium, pH 4.0 (0.04 M phosphate–citrate buffer). All of the tested species proved to be sensitive to the action of cellobiose lipids (the results are presented in Tables 4.2 and 4.3). The activity assay by the rate of survival in liquid media using the same test cultures of yeast fungi showed no sufficient difference between the activities of the above cellobiose lipids either. The minimum concentrations causing total cell death, calculated on the basis of these experiments, are given in Table 4.4. The cellobiose lipids under study demonstrated similar fungicidal activities against the same yeast cultures. So, their concentration causing total cell death was 0.04–0.05 mg/ml for *Cr. terreus* and 0.16–0.2 mg/ml for

Table 4.3 Cell Viability Under Action of Purified Cellobiose Lipid of *Pseudozyma graminicola* VKM-2938

Yeast Species	Cellobiose Lipid Concentration, mg/ml								
	0	0.01	0.02	0.04	0.07	0.10	0.20	0.30	0.50
	Cell Viability, % of Control								
Cryptococcus terreus VKM Y-2253	100	4	2	0	0	0	0	–	–
Filobasidiella neoformans IGC 3957	100	27	16	4	0	0	0	–	–
Trichosporon asteroides VKM Y-275	100	75	34	17	12	1	–	–	–
Trichosporon asahii VKM Y-803	100	100	81	25	12	1	–	–	–
Trichosporon faecalis VKM Y-2865	100	67	32	21	8	2	–	–	–
Candida viswanathii CBS 4024	100	46	40	29	–	3	1	–	–
Candida tropicalis RBF-988	100	–	–	–	43	13	2	1	0
Candida parapsilosis VKM Y-58	100	–	–	–	78	25	6	2	0
Candida glabrata CBS 138	100	100	–	–	–	43	30	23	2
Candida albicans JCM 1542	100	100	–	–	–	21	2	1	0
Clavispora lusitaniae IGC 2705	100	100	–	–	–	4	1	1	0
Saccharomyces cerevisiae VKM Y-1173	100	100	–	–	–	14	5	3	0

The cells of test cultures were treated in 0.04 M citrate–phosphate buffer, pH 4.0, at room temperature for 30 min.

Table 4.4 The Concentrations of Cellobiose Lipids of *Cryptococcus humicola* 9-6 and *Pseudozyma spp.* Causing Complete Cell Death After 30-min Treatment at pH 4.0

Test Cultures	Concentration, mg/ml	
	Cellobiose Lipid of *Cryptococcus humicola*	Cellobiose Lipid of *Pseudozyma* spp.
Basidiomycetes		
Cryptococcus terreus VKM Y-2253	0.02	0.02
Filobasidiella neoformans IGC 3957	0.02	0.02
Trichosporon asahii VKM Y-803	0.05	0.07
Trichosporon faecalis VKM Y-2865	0.14	0.14
Ascomycetes		
Candida albicans JCM 1542	0.15	0.15
Candida glabrata CBS 138	0.3	0.3
Clavispora lusitaniae IGC 2705	0.15	0.1
Candida parapsilosis VKM Y-58	0.23	0.2
Candida tropicalis RBF-988	0.5	0.2
Candida viswanathii CBS 4024	0.15	0.1
Saccharomyces cerevisiae VKM Y-1173	0.3	0.3

Source: *Kulakovskaya et al. (2009).*

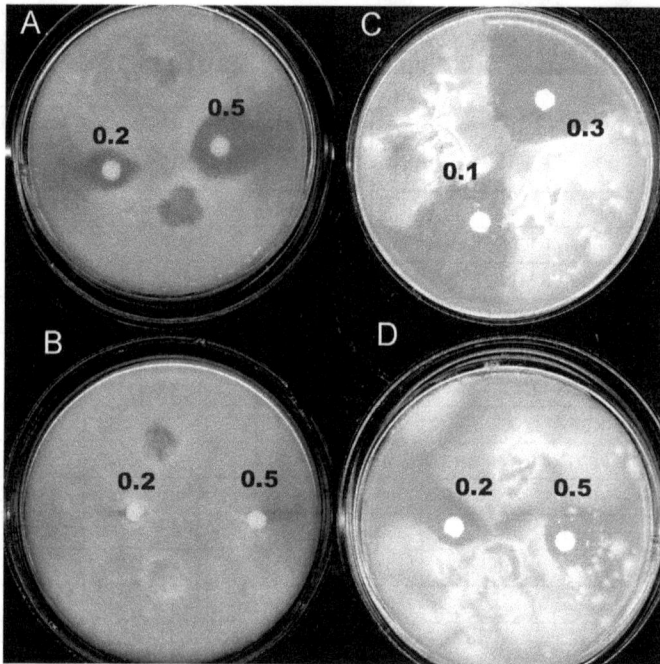

Figure 4.5 The inhibition of fungi growth by purified cellobiose lipids of Pseudozyma fusiformata VKM Y-2821 (A, C) and Cryptococcus humicola 9-6 (B, D) at pH 4.0. The numbers indicate the amount of cellobiose lipids (mg per disc). Test cultures: Sclerotinia sclerotiorum (A, B) and Mucor mucedo VKM F-1355 (C, D).

C. albicans, that is, there was no substantial difference in the fungicidal activity against the above yeasts between glycolipids with O-substituents of different structure. Basidiomycetous yeasts are most sensitive to cellobiose lipids compared to ascomycetes: the cells of cryptococcosis pathogen F. neoformans almost completely died after 30-min incubation with 0.02 mg/ml of the glycolipid. The same effect for ascomycetous yeasts, including candidiasis pathogens, was achieved at eightfold higher glycolipid concentrations. It is still an open question what peculiarities of the cell envelope or membrane structure determine such difference.

The growth of the tested mycelial fungi was inhibited by the amounts of glycolipids similar to those suppressing the growth of asco-mycetous yeasts. The cellobiose lipid of Ps. fusiformata more effec-tively inhibited the growth of fungi (including the phytopathogenic Phomopsis helianthi and Sclerotinia sclerotiorum) compared to the gly-colipid of Cr. humicola (Figure 4.5; Table 4.5).

Table 4.5 The Inhibition of Growth of Fungi by Purified Cellobiose Lipids (Glucose-Peptone Agar, pH 4.0)					
Cellobiose Lipid	Amount, mg per Disc	Growth Inhibition Zone Diameters, mm			
		Sclerotinia sclerotiorum	*Phomopsis helianthi*	*Mortierella isabellina* VKM F-526	*Mucor mucedo* VKM F-1355
Pseudozyma fusiformata	0.1	15	15	12	15
	0.3	20	20	15	20
Cryptococcus humicola	0.2	8	8	0	10
	0.5	12	12	0	15

For bacteria, we have tested *Acetobacter xylinum* VKM B-880, which is able to grow at pH 4.0. Growth inhibition zones were not observed even in the presence of 0.5 mg of cellobiose lipids per disc. Previously it was reported that many bacteria tested by the "culture-to-culture" method demonstrated insensitivity to the glyco-lipid of *Cr. humicola* (Golubev and Shabalin, 1994).

The yeast *Ps. flocculosa* known for its activity against the mildew pathogen *Sphaerotheca fuliginea* (Avis and Belanger, 2002) and phyto-pathogenic fungus *Phomopsis* sp. Cheng et al. (2003) produces a cellobiose lipid with several O-substituents in the cellobiose residue (see Chapter 1, Fig. 1.1) (Mimee et al., 2005). This glycolipid, the so-called flocculosin, is active against a wide range of yeast fungi, including those pathogenic for plants, animals, and humans (Avis and Belanger, 2001; Cheng et al., 2003, Mimee et al., 2005). The minimum inhibitory concentration (MIC) of this cellobiose lipid at pH 5.0 for *C. albicans* ATCC-18804 was 0.025 mg/ml and the total growth inhibition was observed at 0.05−0.1 mg/ml (Mimee et al., 2009a), which is close to our data obtained with another cellobiose lipids. The same article also presents the values of MICs for some species of both Gram-negative and Gram-positive bacteria determined at pH 5.0. MIC of most of Gram-positive bacteria is in the range of 0.01−0.04 mg/ml, while in Gram-negative species it is higher by an order of magnitude. The authors explain this effect by different membrane structures of Gram-positive and Gram-negative bacteria. Moreover, they believe that flocculosin, even though having a fungicidal effect on yeasts, exerts only a bacterio-static effect on the cells of *Staphylococcus aureus*. It is important to note that the glycolipid was not toxic for the human cell lines used by the authors even at the 10- and 100-fold higher concentrations than those inhibiting the growth of *C. albicans* (Mimee et al., 2005).

So, cellobiose lipids may be considered as promising natural compounds for the development of novel fungicides for agriculture and medicine.

At the same time, the application of glycolipid producers for plant colonization showed that other factors may be involved in protection against pathogens. Four phylogenetically-related species, *Ps. antarctica*, *Ps. flocculosa*, *Ps. fusiformata*, and *Ps. rugulosa* were tested as a colonization agents on both healthy and powdery mildew-infected leaves (Clément-Mathieu et al., 2008). Only *Ps. flocculosa* antagonized powdery mildew colonies (Clément-Mathieu et al., 2008). On powdery mildew-infected cucumber leaves, *P. flocculosa* induced a complete collapse of the pathogen colonies, but glycolipid production, as studied by *cyp1* gene expression, was still comparable to that of controls (Hammami et al., 2011). These results suggest that production of glycolipids, and more specifically flocculosin or ustilagic acid, is not the sole factor dictating biocontrol activity among *Pseudozyma* spp. (Clément-Mathieu et al., 2008; Hammami et al., 2011).

4.1.3 Antifungal Activities of Natural Cellobiose Lipids and Their Synthetic Derivatives

Comparative assessment of the antifungal activity of cellobiose lipids was performed both in cellobiose lipids purified from the culture liquid of *Cr. humicola* and *Ps. fusiformata*, deacylated derivatives of cellobiose lipids, containing no O-substituents in the cellobiose residue, the synthetic cellobiose lipid, 16-(β-cellobiosyloxy)-hexadecanoic acid (Kulakovskaya et al., 2009) and 16-hydroxyhexadecanoic acid. The structures of the compounds and the results of antifungal activities testing are shown in Figure 4.3 and Table 4.6, respectively. The natural cellobiose lipids were equally active against the three selected cultures. Fully O-deacylated analogues, namely 16-(β-cellobiosyloxy)-2-hydroxy-hexadecanoic acid, and the synthetic compound of 16-(β-cellobiosyloxy)-hexadecanoic acid, do not inhibit the growth of *F. neoformans* and *S. cerevisiae*, while 16-(β-cellobiosyloxy)-2,15-dihydroxyhexadecanoic acid inhibits the growth of both test cultures but at higher concentrations than cellobiose lipids of *Cr. humicola* and *Ps. fusiformata*. The amide of 16-(β-cellobiosyloxy)-2,15-dihydroxyhexadecanoic acid and 16-hydroxyhexadecanoic acid (ICN, USA) showed no antifungal activity. The simplest cellobiose lipid providing antifungal activity consists of cellobiose residue and fatty acid residue with three hydroxyl groups

Table 4.6 Growth Inhibition Zones Diameters (mm) Under Action of Natural Cellobiose Lipids, Their Deacylated Derivatives, Synthetic 16-(β-Cellobiosyloxy)-Hexadecanoic acid and 16-Hydroxyhexadecanoic Acid at pH 4.0

Compound (the Number of Compound in Figure 4.3 is Given in Parenthesis)	Test Cultures		
	Saccharomyces cerevisiae VKM Y-1173	*Cryptococcus terreus* VKM Y-2253	*Filobasidiella neoformans* IGC 3957
Cellobiose lipid of *Pseudozyma* spp. (1)	5 (0.2)	5 (0.01)	5 (0.01)
Cellobiose lipid of *Cryptococcus humicola* (2)	5 (0.2)	5 (0.01)	5 (0.01)
2,15-Dihydroxy-(β-cellobiosyloxy)-hexadecanoic acid	5 (0.7)	5 (0.7)	5 (0.7)
15-Hydroxy-cellobiosyloxy-hexadecanoic acid	0 (0.8)	0 (0.8)	0 (0.8)
2,15-Dihydroxy-(β-cellobiosyloxy)-hexadecanoic acid acetamide	0 (1)	0 (1)	0 (1)
16-(β-Cellobiosyloxy)-hexadecanoic acid	0 (1)	0 (1)	0 (1)
16-Hydroxyhexadecanoic acid	0 (1)	0 (1)	0 (1)
The amounts of the compounds (mg per disc) are given in parenthesis.			

(Figure 4.3; Table 4.6). Cellobiose lipids with O-substituents showed higher activity. Thus, the structures of both the carbohydrate part and fatty acid aglycon moiety are important for the fungicidal activity of cellobiose lipids.

The presence of such activity in the natural detergents like extracellular glycolipids of yeast fungi is not surprising. The antifungal effect of fatty acids has been reported earlier (Hou and Forman, 2000; Avis and Belanger, 2001; Graner et al., 2003). The antifungal activity is characteristic of the fatty acids with double bonds or with more than 2−3 hydroxyl groups (Niwano et al., 1984; Hou and Forman, 2000; Avis and Belanger, 2001; Graner et al., 2003; Carballeira et al., 2005), or for the fatty acids with a relatively short (C_{9-12}) hydrocarbon chain (Kabara and Vrable, 1977; Bergsson et al., 2001). It should be taken into account that the many detergents, including the known SDS, CTAB (cetyltrimethylammonium bromide), etc., possess antibiotic activities, but their antifungal activity was observed at concentration 0.5−1 M (Kodedova et al., 2011; Kulakovskaya et al., 2011). The antifungal activity of cellobiose lipids is much higher.

We compared the fungicidal activities of *Cr. humicola* cellobiose lipid, fluconazole and commercial preparation of sophorolipids (Sopholiance) against *F. neoformans* IGC 3957 with the standard technique of MIC assay in immunoassay plates at pH 4.0. The MIC of cellobiose lipid was 0.012 mg/ml, of fluconazole it was 0.006 mg/ml while that of sophorolipids was significantly higher (2.4 mg/ml).

4.2 MEMBRANE-DAMAGING ACTIVITY OF CELLOBIOSE LIPIDS

The discovery of ATP release from target cells treated with the cellobiose lipids of *Cr. humicola* (Puchkov et al., 2001) suggested that they possessed a membrane-damaging activity. This suggestion was confirmed by many physicochemical methods (Puchkov et al., 2001, 2002).

We have studied the ATP leakage from yeast cells treated with the cellobiose lipids in more detail, because its measurement proved to be a convenient method for rapid quantitative assessment of fungicidal activity (Kulakovskaya et al., 2003). Figure 4.6 shows time dependence of the ATP release from *C. albicans* cells treated with cellobiose lipids of *Cr. humicola*. The cellobiose lipids of *Pseudozyma* showed similar time dependence of ATP leakage (Kulakovskaya et al., 2003). Considerable ATP leakage occurred after 5-min incubation and reached the maximum in 15 min. ATP leakage was not observed on addition of appropriate amounts of methanol (the experiments were performed with methanol solutions of cellobiose lipids) as a control.

The increase in the temperature of incubation resulted in enhancing of ATP leakage (Figure 4.7), probably due to the well-known property

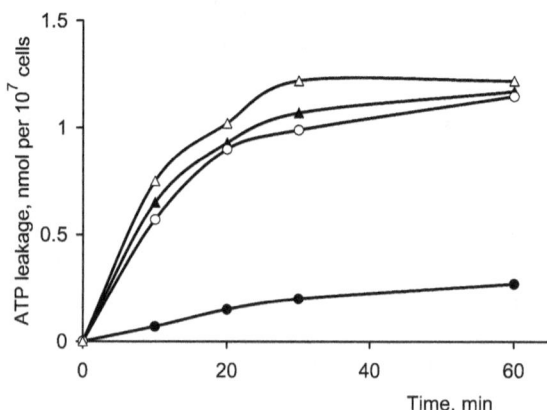

Figure 4.6 The time dependence of ATP leakage from the cells of Candida albicans JCM 1542 treated with cellobiose lipid of Cryptococcus humicola 9-6 at pH 4.0 and 20°C. Cellobiose lipid concentrations (mg/ml): 0.06 (●), 0.09 (o), 0.13 (▲), 0.15 (△).

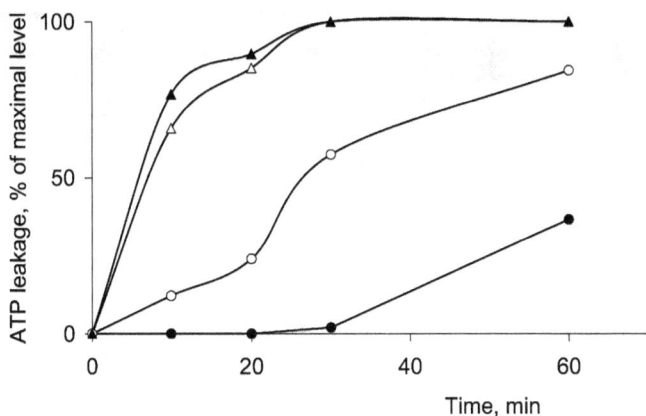

Figure 4.7 The temperature dependence of ATP leakage from the cells of Cryptococcus terreus VKM Y-2253 treated with cellobiose lipid of Cryptococcus humicola 9-6: 0°C (●), 10°C (o), 20°C (Δ), 30°C (▲). Cellobiose lipid concentration was 0.13 mg/ml, pH 4.0.

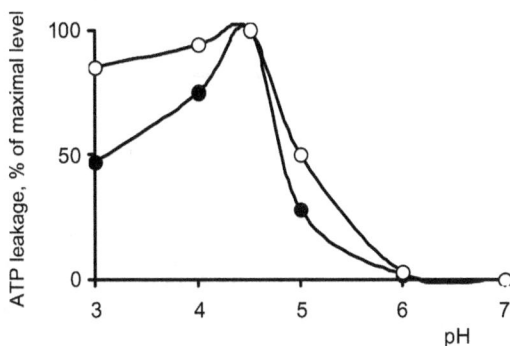

Figure 4.8 The pH dependence of ATP leakage from the cells of Cryptococcus terreus VKM Y-2253 treated with cellobiose lipids (0.06 mg/ml) of Cryptococcus humicola 9-6 (●) and Pseudozyma fusiformata VKM Y-2821 (o) at 20°C for 30 min.

of lipid bilayers: the increase in their fluidity at higher temperatures facilitating the incorporation of detergents into the bilayer.

The maximum ATP leakage was observed at pH 4.0–4.5 (Figure 4.8). It is in agreement with the pH optimum for antifungal activity tested by the "culture-to-culture" method (Golubev and Shabalin, 1994; Golubev et al., 2001). The antifungal activity of cellobiose lipid of *Ps. flocculosa* shows similar pH dependence (Mimee et al., 2005, 2009a).

Table 4.7 The Concentrations of Cellobiose Lipids Causing Loss of Cell Viability and Maximal ATP Leakage

Cellobiose Lipid	Cell Viability ~1% of Control		Maximal ATP Leakage	
	Cryptococcus terreus VKM Y-2253	*Candida albicans* JCM 1542	*Cryptococcus terreus* BKM Y-2253	*Candida albicans* JCM 1542
	Cellobiose Lipid Concentration, mg/ml			
Pseudozyma fusiformata	0.02	0.20	0.06	0.10
Cryptococcus humicola	0.02	0.16	0.05	0.12

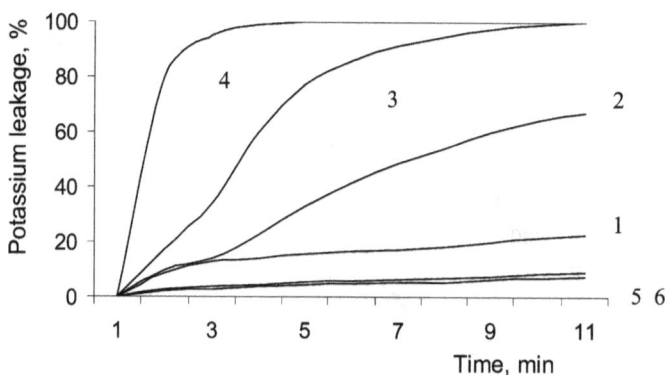

Figure 4.9 Potassium leakage from the cells of Saccharomyces cerevisiae VKM Y-1173 treated with cellobiose lipid of Pseudozyma fusiformata VKM Y-2821 at pH 4.0. (1), 0.14 mg/ml of glycolipid; (2), 0.22 mg/ml of glycolipid; (3), 0.29 mg/ml of glycolipid; (4), 20 μM Ag⁺; (5), 2% methanol; (6), 2% of ethanol.

We determined the dependence of ATP leakage on cellobiose lipids concentration for some test cultures (Table 4.7). The concentrations inducing the maximum ATP leakage were near to those resulting in the total cell death of test cultures under study.

The leakage of potassium ions from target cells is another evidence for the membrane-damaging effect of cellobiose lipids of *Cr. humicola* and *Ps. fusiformata* (Kulakovskaya et al., 2008, 2011). The cellobiose lipids caused concentration-dependent K^+ release from the cells of test cultures (Figure 4.9). At enhanced concentrations of the fungicide, the K^+ release was intensified and reached the same level as under the influence of Ag^+. The K^+ release begins in the very first minutes of exposure, reflecting the ability of cellobiose lipids to impair membrane integrity at the initial stage of interaction with target cells. The leakage of potassium ions was observed also when spheroplasts were treated

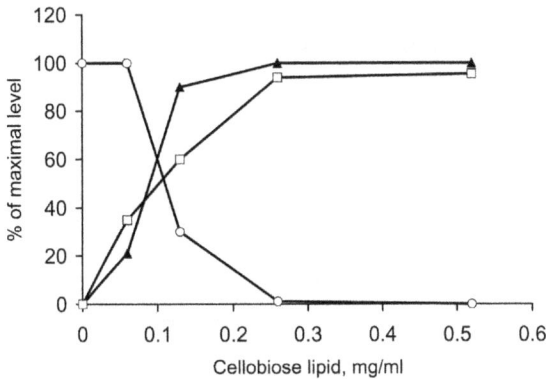

Figure 4.10 The cells viability (o), ATP leakage (▲), and staining with bromocresol purple (□) of the cells of Cryptococcus terreus VKM Y-2253 treated with cellobiose lipid of Pseudozyma fusiformata at pH 4.0 and 20°C for 30 min.

with cellobiose lipids (Vagabov et al., 2008). These data demonstrate that the cell envelope makes no significant contribution to the interaction between cellobiose lipid and target cells. The release of potassium ions from the cells of *C. albicans* treated with cellobiose lipid of *Ps. flocculosa* was demonstrated and interpreted as an evidence for the membrane-damaging activity (Mimee et al., 2009a).

We have also performed other tests to confirm the ability of cellobiose lipids to damage the membranes. The cells of *Cr. terreus* VKM Y-2253, before the treatment with this compound and after the treatment with its different concentrations, were strained with bromocresol purple. This dye does not penetrate undamaged cells and has been used by other authors to prove the membrane-damaging activity of mycocines (Kurzweilova and Sigler, 1993). The increase in staining intensity and ATP leakage and decrease in cell survival showed similar dependence on the concentration of cellobiose lipid of *Ps. fusiformata* (Figure 4.10).

One of the criteria of plasma membrane integrity for *S. cerevisiae* is acidification of the medium in the presence of glucose due to proton release by the plasma membrane H^+-ATPase (Petrov et al., 1992). If the cytoplasmic membrane is damaged, its proton gradient cannot be maintained. The treatment of *S. cerevisiae* cells with cellobiose lipid resulted in the loss of this activity (Figure 4.11). Similar data were obtained in the experiments with *C. albicans* (Mimee et al., 2009a,b).

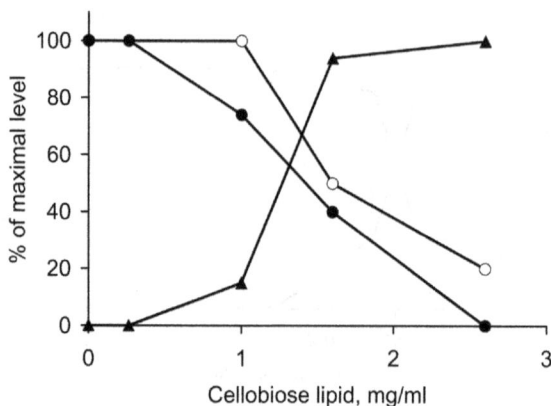

Figure 4.11 The cells viability (o), ATP leakage (▲), and medium acidification (●) of the cells of Saccharomyces cerevisiae VKM Y-1173 treated with cellobiose lipid of Pseudozyma fusiformata at pH 4.0 and 20°C for 30 min.

Cellobiose lipid of Ps. flocculosa caused great ultrastructural changes in the target cells revealed by electron microscopy: damage to the cytoplasmic membrane, formation of numerous membrane vesicles in the cytoplasm, and damage to the nuclear envelope and the mitochondrial membranes (Mimee et al., 2009a). The authors believe that cellobiose lipids have no specific target and that the observed damage is associated with their detergent properties. The absence of specific target molecules is also confirmed by the fact that cellobiose lipids destroy liposomes of different composition (Puchkov et al., 2002).

It is interesting to note that the fungicidal effect of cellobiose lipids is blocked in the presence of 100 μM $CaCl_2$ (Mimee et al., 2009a). It is probably due to the formation of calcium salts that are poorly soluble even in buffer solutions, similar to the formation of calcium and magnesium salts of fatty acids when using soap in hard water.

In spite of a rather long period of studying the membrane-damaging activity of cellobiose lipids, it is still unclear what factors determine a different sensitivity of yeast species from different taxa and what is the target of these compounds in the cytoplasmic membrane. Mimee et al. (2009a) compared the effects of flocculosin and amphotericin B, the well-known membrane-damaging compound, on acidification of the medium by target cells and on the release of potassium ions and found that the effect of amphotericin B was much more rapid. It is known that amphotericin B is bound to ergosterol through hydrogen bonds and forms a specific pore, which results in perturbation of the proton gradient and depolarization of the cytoplasmic membrane (Brajtburg et al., 1990). As amphotericin and flocculosin differ in some parameters of interaction with the membrane, it has been suggested that ergosterol cannot be a target of flocculosin in the cytoplasmic membrane; at least, it is not a specific target, because the oomycetes not synthesizing this lipid are nevertheless sensitive to flocculosin (Mimee et al., 2009a).

We have revealed the lower sensitivity to the *Cr. humicola* cellobiose lipid in the *S. cerevisiae* mutant deficient in ergosterol synthesis and the higher sensitivity in the mutant deficient in sphingomyelin synthesis.

Not only various yeast species and strains have different sensitivity to cellobiose lipids but also the same species/strain show differences in sensitivity to cellobiose lipids depending on the peculiarities of cultivation. Cultivation on ethanol-containing media causes considerable changes in the composition and properties of the cytoplasmic membrane, including the higher amounts of ergosterol and unsaturated fatty acids in phospholipids, which results in reduced fluidity of cytoplasmic membranes (Susan et al., 1978; Beaven et al., 1982; Mishra and Prasad, 1989; Walker-Caprioglio et al., 1990; Herve et al., 1994; Kubota et al., 2004). The cells of *S. cerevisiae* grown on ethanol have lower sensitivity to cellobiose lipid of *Cr. humicola* when compared to the cells grown on glucose (Trilisenko et al., 2012) (Table 4.8).

Table 4.8 Cell Viability of *Saccharomyces cerevisiae* VKM Y-1173 Grown in the Media with Glucose and Ethanol Under Treatment with Cellobiose Lipid of *Cryptococcus humicola* 9-6

Cellobiose Lipid Concentration, mg/ml	Viability, %	
	The Cells Grown on Glucose	The Cells Grown on Ethanol
0	100	100
0.025	78	–
0.034	10	–
0.050	4	100
0.100	0.2	14
0.200	0.15	2
0.400	0.01	1
0.800	–	0.05
Source: Trilisenko et al. (2012).		

On the face of it, the data on reduced sensitivity to cellobiose lipids in the mutant with the lower content of ergosterol and in the cells grown on ethanol and containing, according to the literature data, more ergosterol, contradict each other. However, it is necessary to take into consideration that the target for the action of cellobiose lipids on the cytoplasmic membrane is yet unknown. The mechanism of their incorporation into the membrane is far from being finally described even for the fungicides with the membrane-damaging effect such as amphotericin B that have been extensively used for a long time (Mimee et al., 2009a). It is not improbable that no specific target will be revealed for cellobiose lipids; however, the increase in membrane viscosity will probably be an obstacle for incorporation of cellobiose lipids.

Glucose induces rapid and significant changes in the membrane of *S. cerevisiae* (Permyakov et al., 2012). We have checked how glucose influences the sensitivity to cellobiose lipids. The sensitivity of *S. cerevisiae* cells to cellobiose lipid increased on addition of glucose to the incubation medium. This fact has been demonstrated by both ATP leakage (Table 4.9) and potassium ions leakage (Kulakovskaya et al., 2008a). However, such effect was not observed in the cells grown in the medium with ethanol (Table 4.9).

We also compared the effect of cellobiose lipid on the cells of *S. cerevisiae* with different polyphosphate content. Simultaneously, we compared the effects of silver ions and the known anionic detergent SDS on the above cells. The cells of *S. cerevisiae* that have accumulated polyphosphates are characterized by their enhanced content in the cell wall and the higher negative charge of the cell envelope (Ivanov et al., 1996). The measurement of the release of ATP (Table 4.10) and potassium ions (Kulakovskaya et al., 2011) showed that the sensitivity to the cellobiose lipid of *Ps. fusiformata* did not depend on polyphosphate

Table 4.9 The Concentrations of Cellobiose Lipid (mg/ml) of *Cryptococcus humicola* 9-6 Causing the Maximal Leakage of ATP from the Cells *Saccharomyces cerevisiae* VKM Y-1173		
Incubation Media for Treatment with Cellobiose Lipid	The Cells Grown in the Medium with Glucose	The Cells Grown in the Medium with Ethanol
0.4 M phosphate–citrate, pH 4.0	0.6	0.6
0.4 M phosphate–citrate, pH 4.0 with 2% of glucose	0.15	0.6
Source: *Trilisenko et al. (2012).*		

Table 4.10 The Sensitivity of the Cells of *Saccharomyces cerevisiae* VKM Y-1173 with Different Level of Inorganic Polyphosphates to Cellobiose Lipid of *Pseudozyma fusiformata*, SDS, and Ag^{2+}

The Cell Type	Polyphosphate Level, mM Phosphorus/g of Dry Biomass	Concentration Causing ATP Leakage ($\sim$30% of Maximal), mM		
		Cellobiose Lipid	SDS	AgNO$_3$
Control cells	0.51	0.1	0.4	0.005
The cells with enhanced polyphosphate level	1.09	0.1	2.0	0.02
Source: Kulakovskaya et al. (2011).				

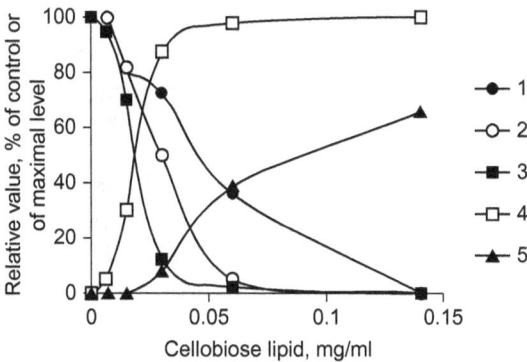

Figure 4.12 The effect of cellobiose lipid of Pseudozyma fusiformata on spheroplasts of Saccharomyces cerevisiae VKM Y-1173: 1, orthophosphate in spheroplasts; 2, polyphosphate in spheroplasts; 3, ATP in spheroplasts; 4, ATP leakage; 5, α-glycosidase leakage.

content, while the sensitivity to SDS and silver ions decreased at higher polyphosphate levels.

Cellobiose lipids can also have an effect on spheroplasts. The treatment of *S. cerevisiae* spheroplasts with the cellobiose lipid of *Ps. fusiformata* was accompanied by the release of ATP, the 50% release of α-glucosidase (the enzyme localized in the cytoplasm), as well as the decline in P$_i$ level and suppression of polyphosphate accumulation (Figure 4.12). The effective concentration of cellobiose lipids for spheroplasts was twice lower than for intact cells. The release of α-glucosidase from the cells was not observed, probably because it was prevented by the cell wall. Spheroplasts are more sensitive to the effect of cellobiose lipid. At a cellobiose lipid concentration reduced to 0.03 mg/ml, ATP release was maintained at the previous level, the release of α-glucosidase was not observed, P_i content decreased, and polyphosphate accumulation was inhibited by 50%. The polyphosphates previously accumulated by spheroplasts are not released into the medium under the effect of glycolipid, while ATP content decreases. If spheroplasts that have accumulated polyphosphates are incubated in the medium lacking phosphate and glucose, the polyphosphate level does not vary for at least 1 h. In the presence of cellobiose lipid, the level of ATP considerably decreased due to its release into the medium, the P_i level also decreased, whereas the polyphosphate level remained close to the control value. The findings indicate that the accumulated polyphosphates are retained in spheroplasts even in the case of partial impairment of the cytoplasmic membrane permeability. The data on the glycolipid effect on spheroplasts (Kulakovskaya et al., 2008a; Vagabov et al., 2008) demonstrate that, although the cell wall partially participates in the interaction between cellobiose lipids and cells, it is not a target of these compounds.

Cellobiose lipids are effective membrane-damaging agents that seem to interact directly with the cytoplasmic membrane by forming pores in it, which results in the loss of low-molecular vital compounds and disturbance of ion gradients at the membrane. If the period of influence is longer, cellobiose lipids can penetrate the cell and damage the membranes of organelles, which results in cell death. The acidic pH optimum for fungicidal effect seems to be associated with the ionization of the carboxylic group.

Owing to the broad spectrum of action, thermal stability, storage stability, the ability to inhibit the growth of species that are pathogenic for humans and animals and cause plant diseases, affect fruit and roots, and impair food quality, cellobiose lipids, and their producers are promising for the development of novel fungicidal preparations for laboratory use, agriculture, and medicine.

4.3 BIOLOGICAL ACTIVITIES OF MELs AND SOPHOROLIPIDS

There is a widely-held view that extracellular glycolipids perform primarily the function of biosurfactants, that is, compounds facilitating solubilization and absorption by the cells of various organic hydrophobic compounds that are present in the medium and can be utilized by microorganisms as growth substrates. A comprehensive review of this role of extracellular glycolipids is presented in Spencer et al. (1979), Lang and Wagner (1987), Rosenberg and Ron (1999), Cameotra and Makkar (2004), Kitamoto et al. (2002), Rodrigues et al. (2006), Langer et al. (2006), Van Bogaert et al. (2007a), Arutchelvi et al. (2008), Bolker et al. (2008), Arutchelvi and Doble (2011), and Van Bogaert and Soetaert (2011).

The antibiotic activity of sophorolipids and MEL has been reported. The growth of Gram-positive bacteria (*Bacillus subtilis* and *Micrococcus luteus*) was suppressed by 3–10 mg/l of MEL or 0.12–0.48 mg/l of sophorolipid (Kitamoto et al., 2002), but higher concentrations (up to 400 mg/l) were needed to suppress the growth of Gram-negative bacteria. Sophorolipids inhibit the growth of Gram-positive bacteria such as *Staphylococcus epidermidis, Staphylococcus aureus, Streptococcus faecium, Propionibacterium acnes,* and *Corynebacterium xerosis* (Hommel et al., 1987; Lang et al., 1989).

Synthesized analogs of sophorolipids were obtained and their antibiotic activities were studied (Azim et al., 2006). These compounds consist of amino acids linked by amide bonds to the carboxyl group of fatty acid residues of sophorolipids. All tested analogs showed antibacterial activity against both Gram-positive and Gram-negative bacteria. Leucine conjugate was the most efficient: the MICs for *Moraxella* sp. and *Escherichia coli* were 0.83 mg/ml and 1.67 mg/ml, respectively. All compounds displayed virus-inactivating activity with 50% effective concentrations below 0.2 mg/ml (Azim et al., 2006).

Some authors disprove the antibacterial activity of sophorolipids and believe that sophorolipids can be used as immunomodulators without any effect on the host microflora (Sleiman et al., 2009).

The data on the fungicidal activities of sophorolipids and MEL should be considered with care, because their effective concentrations are rather high. The MICs against *C. albicans* were above 400 mg/l for MEL-A and MEL-B and above 200 mg/l for the sophorolipid of *C. apicola* (Kitamoto et al., 2002). These concentrations are not much more effective than MIC of the known detergent SPAN-20 (sucrose nonadecanoate).

The antifungal activities of *St. bombicola* sophorolipids and some of their derivatives were tested by the standard MIC assay during the cultivation of test cultures in plates on the medium containing yeast extract and wort at a concentration of 5 mg/ml. The natural sophorolipid mixture suppressed the growth of *C. albicans* and *C. tropicalis* by 30% and 25%, respectively, while some derivatives esterified to the fatty acid residue (ethyl-17-L-[(2′-*O*-β-D-glucopyranosyl-β-D-glucopyranosyl)-oxy-]-cys-9-octadecanoate-6′6″-diacetate and methyl-17-L-[(2′-*O*-β-D-glucopyranosyl-β-D-glucopyranosyl)-oxy-]-cys-9-octadecanoate) were capable of 100% growth suppression of these two species (Gross and Shah, 2005).

Diverse biological activities of sophorolipids and MEL are associated primarily with their surfactant and amphiphilic properties.

For instance, it has been shown that MEL-A substantially increases the efficiency of transfection by cationic liposomes. It is supposed that MEL-A induces the membrane fusion of cationic liposomes and target cells and thereby accelerates the transfer of genetic material (Inoh et al., 2001, 2004).

The biological activities of MEL and sophorolipids were actively studied in the context of their influence on various regulatory processes in mammalian cells. These data were summarized in the reviews (Kitamoto et al., 2002; Cameotra and Makkar, 2004; Arutchelvi and Doble, 2011; Van Bogaert and Soetaert, 2011).

The experiments with cell cultures have shown the anticancer activity of MEL and sophorolipids. MEL can suppress the proliferation of lymphocytes affected by leukemia in cell culture and induce their differentiation (Isoda et al., 1997), inhibit the proliferation of melanoma cells and, at sufficiently high concentrations, induce their apoptosis (Zhao et al., 1999, 2001).

Sophorolipids also suppressed the reproduction of liver carcinoma cells (Chen et al., 2006a,b). The 10 sophorolipid molecules differing in acetylation degree of sophorose, unsaturation degree of hydroxyl fatty acid, and lactonization or ring opening were tested as inhibitors of esophageal cancer cells (Shao et al., 2012). The inhibition of diacetylated lactonic sophorolipid (total inhibition at 30 μg/ml) was stronger than that of monoacetylated lactonic sophorolipid (totally inhibition at 60 μg/ml). The sophorolipid with one double bond in fatty acid part had the strongest cytotoxic effect on two esophageal cancer cells. Acidic sophorolipid showed higher anticancer activity (Shao et al., 2012).

Cytotoxicity of the natural sophorolipid mixture and individual chemical derivatives of these compounds against pancreas carcinoma cells was investigated (Fu et al., 2008). It was found that sophorolipid methyl esters proved to be more toxic for these carcinoma cells compared to the natural sophorolipids. Sophorolipids in the acidic form and the sophorolipid containing two acetate groups and existing in the lactone form demonstrated similar toxicities (the death of 40–50% cells after 24-h treatment at a concentration of 0.5 mg/ml). Under these conditions, the mononuclear cells of peripheral blood were insensitive to all of the used sophorolipids and their derivatives (Fu et al., 2008).

The effects of MEL and sophorolipid on the initiation of nerve ending growth were investigated. The addition of these glycolipids was shown to cause a considerable growth of nerve endings. MEL-A enhances the acetylcholine esterase activity to a level similar to that caused by the nerve tissue growth factor (Isoda et al., 1999). Other authors have shown that MEL increases the level of galactoceramide and the growth of nerve endings in pheochromocytoma cells (Shibahara et al., 2000). In addition, the treatment of the cells of this line with MEL leads to cell cycle interruption in the G1-phase and suppression of cell differentiation (Wakamatsu et al., 2001). The mechanism of MEL action is still unknown.

MELs have a high affinity to the human immunoglobulin G. The maximum binding capacity was shown for MEL-A (Im et al., 2001, 2003). It is evident in favor of potential involvement of this glycolipid in binding of other important regulatory proteins and possible regulation of their activities.

Sophorolipids can modulate inflammatory response and reduce mortality under experimentally induced sepsis in rats (Bluth et al., 2006; Hardin et al., 2007). Sophorolipids reduced the production of nitrogen oxide and cytokins by macrophages *in vitro* (Bluth et al., 2006). It was shown that the survival of rats could be enhanced by 40% at a dose of 5 mg/kg of animal weight, while the toxic effect occurred at 75- to 150-fold higher doses (Hardin et al., 2007).

MELs, similar to sophorolipids, influence the inflammatory process (Morita et al., 2011d). The experiments with mast cells have shown the anti-inflammatory effect of MEL due to exocytosis suppression and the inhibition of antigen-induced secretion of leukotriene C(4) and cytokine TNF-α. This effect is determined by suppression of the activity of several signaling systems, including those related to the increase in Ca^{2+} and MAP kinase concentrations and the activities of other signaling systems. MEL also suppressed the phosphorylation of the receptor protein SNARE, which plays a key role not only in exocytosis but also in the intracellular movement of vesicles.

It was shown that 5% and 10% MEL-A solutions yielded 73% and 91% survival of skin cells treated with the damaging concentrations of SDS. MEL-B possessed similar protective properties (Morita et al., 2009a−c, 2011b−d; Yamamoto et al., 2012). It was close to the protectant properties of natural ceramides used in cosmetics for regeneration of damaged skin. The antioxidant properties of MEL were shown using the model of the peroxide-induced oxidative stress of human fibroblast culture; MEL-C proved to be the best antioxidant (Takahashi et al., 2012). It was also shown that MEL-A at a concentration of 0.001 µg/l stimulated the activity of hair bulb cells (Morita et al., 2010a); hence, it was proposed to use mannosylerythritols as agents for regeneration of damaged hair and stimulation of hair growth (Morita et al., 2010b).

The diacetylated MEL (MEL-A) produced from soybean oil significantly increased the viability of the fibroblast cells over 150% compared with that of control cells (Morita et al., 2010c). The monoacetylated MEL (MEL-B) hardly increased the cell viability. The viability of the fibroblast cells decreased with the addition of more than 1 µg/l of MELs, whereas the cultured human skin cells showed high viability with 5 µg/l of MELs. The papilla cells were dramatically activated with 0.001 µg/l of MEL-A produced from soybean oil: the cell viability reached at 150% compared with that of control cells (Morita et al., 2010c).

Sophorolipids have a spermicide activity and can inactivate the human immunodeficiency virus (Shah et al., 2005).

Sophorolipids and MEL are characterized by structural diversity, including different degrees of acetylation and lengths of fatty acid chains. The question about the interrelationship between the structures and effects of these compounds on animal and humans cells is still far from being solved.

Numerous studies have demonstrated that sophorolipids and MEL are not toxic for noncancer human cells (Ikeda et al., 1986 a,b;

Kitamoto et al., 2002). Nontoxicity of cellobiose lipids has also been shown for some cell lines (Mimee et al., 2005). Hence, they may be considered promising for application in cosmetology and medicine as natural detergents, immunomodulators, liposomal carriers for drug delivery, while cellobiose lipids may be considered as prospective fungicidal compounds.

4.4 THE BIOLOGICAL ACTIVITIES OF RARE FUNGAL GLYCOLIPIDS

Roselipins produced by *Gliocladium roseum* (Tabata et al., 1999) are diacylglycerol acyltransferase inhibitors (Tomoda et al., 1999). The inhibitors of this enzyme are considered as potential drugs for obesity (Cases et al., 1998). Emmiguacines are of interest as inhibitors of the replication of influenza type A virus (Boros et al., 2002). The glycolipid of *D. spathularia* exhibits antifungal and antibacterial activities (Stadler et al., 2012).

4.5 THE ROLE OF EXTRACELLULAR GLYCOLIPID FOR YEAST PRODUCERS

The fact that extracellular glycolipids in the first place perform the role of biosurfactants, that is, compounds facilitating solubilization and absorption by the cells of various organic hydrophobic compounds, which are present in the environment and can be utilized by microorganisms as their growth substrates, is generally accepted in the literature and concerns not only the compounds produced by yeast but also the glycolipids secreted by bacteria. This role of extracellular glycolipids is comprehensively reviewed in Spencer et al. (1979), Lang and Wagner (1987), Rosenberg and Ron (1999), Cameotra and Makkar (2004), Arutchelvi and Doble (2011), and Van Bogaert and Soetaert (2011). It is attributed primarily to sophorolipids and MELs, because the cultivation of sophorolipid- and MEL-producing yeasts on the media containing hydrophobic carbon sources considerably increases the production of these compounds (Arutchelvi and Doble, 2011; Van Bogaert and Soetaert, 2011).

Sophorolipids intensify the microbial biodegradation of low-soluble compounds in model incubation systems and soil suspension (Schippers et al., 2000). It has been shown that sophorolipids are more effectively synthesized on the media containing low-soluble substrates, for example, plant oils or alkanes (Daniel et al., 1999; Otto et al., 1999).

In view of the fact that sophorolipids and MELs are produced in large amounts under nitrogen starvation and excessive carbon, it is suggested that these are secondary metabolites formed by fungal cells as an extracellular reserve compound (Van Bogaert et al., 2007). It seems to be one of many quite probable functions of extracellular glycolipids, especially in view of the fact that their producers possess not only the enzyme systems for biosynthesis of these compounds but also the respective systems for their catabolism.

The role of extracellular glycolipids in the vital activity of producers seems to be as follows:

— these biosurfactants facilitate solubilization and absorption of hydrophobic substrates;
— they are futile compounds, secondary metabolites, playing the role of an extracellular reserve of carbon sources; and
— they possess antibiotic activity and give their producers an advantage in competition for the natural ecological niches, such as soil, nectaries, and leaves of plants.

CHAPTER 5

Metabolism of Yeast Extracellular Glycolipids

5.1 BIOSYNTHESIS OF EXTRACELLULAR GLYCOLIPIDS

The biosyntheses of extracellular yeast glycolipids have much in common. First, it is the hydroxylation of fatty acids. It is followed by sequential transfer of carbohydrate residues to the fatty acid residue by glycosyltransferases and, at the final step, attachment of O-substituents, acetate groups (by acetyltransferases), and short-chain fatty acid residues (by acyltranferases). Special enzymes are responsible for the biosynthesis of specific short-chain fatty acids. The sequencing and annotation of the complete genome sequences of *U. maydis* (Kämper et al., 2006), *Ps. antarctica* (Morita et al., 2013), and *St. bombicola* (Van Bogaert et al., 2013) have led to clarification of the biosynthetic pathways of their glycolipids. The major enzymes of glycolipid biosynthesis have been revealed. The genes encoding the enzymes of biosynthesis of MELs (Hewald et al., 2006), cellobiose lipids (Bolker et al., 2008; Teichmann et al., 2007, 2011a,b), and sophorolipids (Van Bogaert et al., 2013) are chromosomal genes found in coregulated clusters. It is typical of the fungal genes involved in secondary metabolism.

5.1.1 Biosynthesis of MEL

MEL biosynthesis, although also involving glycosyltransferases, acyltransferases, and acetyltransferases, is performed by its own specific set of enzymes not related to cellobiose lipid biosynthesis. This is true even in fungi such as *U. maydis* capable of producing both types of glycolipids. Some aspects of MEL biosynthesis in *Ps. antarctica* (*C. antarctica*) have been studied by Kitamoto et al. (1993, 1995, 1998, 1999). If the fatty acids used as a substrate have chain lengths of 12−18 carbon atoms, the fatty acids of MEL are found to be shorter by 2−6 carbon atoms. Cerulenin, the strong inhibitor of fatty acid synthesis *de novo*, had a weak effect on MEL production and fatty acid composition (Kitamoto et al., 1995), while the addition of 2-bromooctanoic acid, inhibiting β-oxidation of fatty acids, drastically reduced MEL production, and the degree of inhibition increased together with the number of carbon atoms in the fatty acids used as a substrate (Kitamoto et al., 1998). More detailed characterization of fatty acids within MEL has shown that they are mainly the intermediate products of β-oxidation of the hydrophobic carbon sources (Kitamoto et al., 1999).

The glycosyltransferase-encoding gene *emt1* of *U. maydis* was the first of the identified genes of MEL biosynthesis (Hewald et al., 2005). The enzyme Emt1 transfers the mannose residue from Guanosine diphosphate (GDP)-mannose to erythritol. The transcription of the *emt1* gene is strongly induced under nitrogen starvation (Hewald et al., 2005). Later on, other genes encoding the enzymes of MEL synthesis in *U. maydis* were revealed (Hewald et al., 2006). The gene cluster for MEL biosynthesis comprises the acetyltransferase gene *mat1*, the *mmf1* gene encoding the putative transport protein responsible for MEL secretion, the acyltransferase-encoding *mac1* and *mac2*, and the glycosyltransferase-encoding *emt1* (Table 5.1). The scheme of the biosynthetic pathway of MEL (Figure 5.1) was proposed and confirmed by the study of mutant MEL derivatives (Hewald et al., 2006).

The *PaEMT1* gene encoding the erythritol/mannose transferase of *Ps. antarctica* was identified and its high identity with the *emt1* of *U. maydis* was detected (Morita et al., 2010a). The genome sequence of *Ps. antarctica* was determined and annotated and then the gene cluster containing five genes responsible for MEL biosynthesis and secretion was revealed. The *PaEMT1*, *PaMac1*, *PaMac2*, *PaMMF1*, and *PaMat1* genes in *Ps. antarctica* show the high level of identity (73%, 59%, 52%, 75%, and 53%, respectively) to the corresponding genes of *U. maydis* (Morita et al., 2013). This fact indicates similarities between the biosynthetic pathways of MEL in both organisms.

5.1.2 Biosynthesis of Cellobiose Lipids

U. maydis secretes a number of secondary metabolites (Bölker et al., 2008). The secretion of cellobiose lipids occurs under nitrogen deficiency (Haskins, 1950). The gene encoding the fatty acid hydroxylase responsible for the biosynthesis of hydroxypalmitic acid was the first of those revealed (Hewald et al., 2005). It has been shown for the mutants of *Ps. flocculosa* with different levels of antifungal activity that this activity is associated with several chromosomal genes (Cheng et al., 2003).

Table 5.1 The Gene Clusters for Biosynthesis of Extracellular Yeast Glycolipids

Glycolipid	Species	Gene	Function	References
Mannosylerythritol lipid	*Ustilago maydis*	*mat1*	Acetyltransferase	Hewald et al. (2006)
		mmf1	A member of facilitator family	
		mac1	Putative acyltransferase	
		emt1	Glycosyltransferase	
		mac2	Putative acyltransferase	
Cellobiose lipid (ustilagic acid)	*Ustilago maydis*	*rua1*	Regulation of gene cluster	Teichmann et al. (2007, 2011a,b)
		cyp2	ω−1-Hydroxylase of palmitic fatty acid	
		fas2	Synthesis of short- and/or long-chain fatty acid	
		atr1	Export of glycolipid	
		uat1	Acetyl-/acyltransferase	
		cyp1	ω-Hydroxylase of palmitic acid	
		uat2	Acetyl-/acyltransferase	
		orf1	Unknown	
		uhd1	Hydroxylase of short-chain fatty acid	
		ugt1	Glycosyltransferase	
		orf2	Unknown	
		ahd1	α-Hydroxylase of palmitic acid	
Cellobiose lipid (flocculosin)	*Pseudozyma flocculosa*	*rfl1*	Regulation of gene cluster	Teichmann et al. (2011a,b)
		fhd1	Hydroxylase of short-chain fatty acid	
		fas2	Synthesis of short- and/or long-chain fatty acid	
		atr1	Export of glycolipid	
		fgt1	Glycosyltransferase	
		fat3	Acetyl-/acyltransferase	
		cyp2	ω−1-Hydroxylase of palmitic acid	
		cyp1	ω-Hydroxylase of palmitic acid	
		fat1	Acetyl-/acyltransferase	
		fat2	Acetyl-/acyltransferase	
		orf1	Unknown	
Sophorolipid	*Starmerella bombicola*	*adn*	Putative alcohol dehydrogenase	Van Bogaert et al. (2013)
		ugtB1	Glycosyltransferase II	
		mdr	Sophorolipid transporter	
		at	Acetyltransferase	
		ugtA1	Glycosyltransferase I	
		cyp52m1	Cytochrome P450 monooxygenase (fatty acid hydroxylase)	
		orf	Unknown	

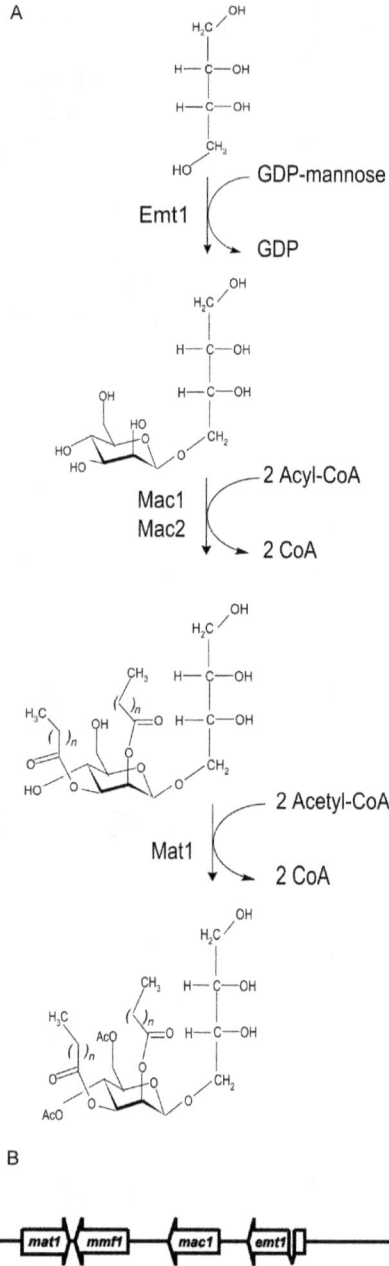

Figure 5.1 MEL biosynthetic pathway (A) and MEL biosynthesis gene cluster (B) (Hewald et al., 2006). A: Emt1—glycosyltransferase, Mac1 and Mac2—acyltransferases; Mat1—acetyltransferase. B: mat1—acetyltrans-ferase gene, mmf1—gene encode a putative MEL transporter, mac1 and mac2—acyltransferase genes, emt1—glucosyltranferase gene.

Ustilago maydis:

Pseudozyma flocculosa:

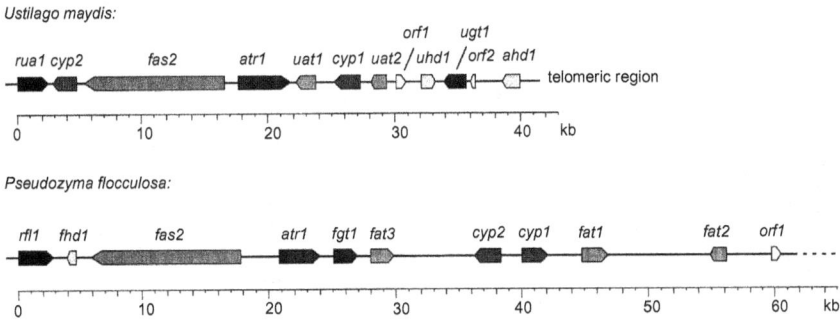

Figure 5.2 Ustilagic acid and flocculosin biosynthesis gene clusters (Teichmann et al., 2011b). The genes and their functions are indicated in Table 5.1 (By permission of John Wiley and Sons).

The genes encoding the enzymes of biosynthetic pathways of cellobiose lipids have been identified for *U. maydis* and *Ps. flocculosa* (Marchand et al., 2009; Hammami et al., 2010; Teichmann et al., 2007, 2010, 2011a,b). Figure 5.2 shows the structure of the gene clusters of *Ps. flocculosa* and *U. maydis* (Teichmann et al., 2011b). There is a high homology between the respective genes of both organisms (Teichmann et al., 2011b). Table 5.1 presents the functions of the proteins encoded by these genes (Teichmann et al., 2007, 2010, 2011a,b). The scheme of cellobiose lipid biosynthesis (Figure 5.3) for *Ps. flocculosa* and *U. maydis* has been postulated on the basis of testing the structures of glycolipids and their precursors in the mutant strains with deletions in the above genes (Teichmann et al., 2007, 2010, 2011a,b).

The biosynthesis of cellobiose lipids begins with hydroxylation of palmitic acid at position 16 by the enzyme cytochrome P450 monooxygenase Cyp1, followed by hydroxylation at position 15 by P450 monooxygenase Cyp2. The homologs of the *cyp1* gene of *U. maydis*, which encodes the cytochrome P450 monooxygenase specific for the biosynthetic pathway of cellobiose lipids, have been found in *Ps. flocculosa* and *Ps. fusiformata* but not in other members of the genus *Pseudozyma* showing no cellobiose lipid secretion (Marchand et al., 2009). The mutants in this gene do not produce cellobiose lipids (Marchand et al., 2009). Then, uridin diphosphate (UDP)-dependent glycosyltransferases Ugt1/Fgt1 transfer glucose residues one by one until the simplest cellobiose lipid without any O-substituents is formed. The subsequent reactions are catalyzed by acetyl- and acyltransferases. Specific short-chain fatty acid hydroxylases have been revealed, namely, the Uhd1 hydroxylase of *U. maydis* and the FnD1

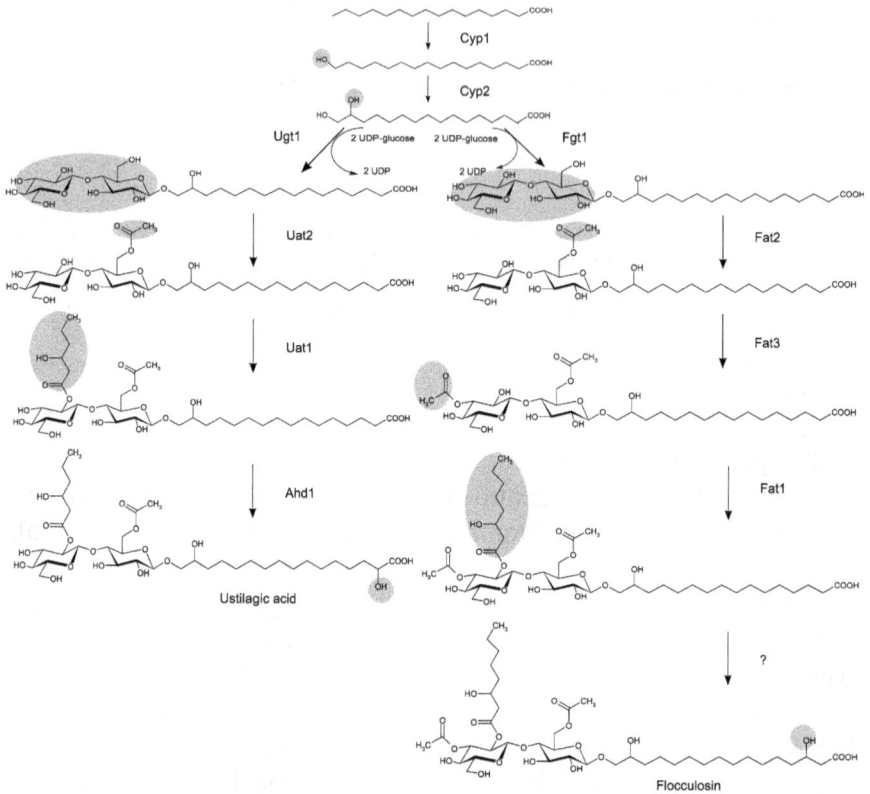

Figure 5.3 Biosynthetic pathways for cellobiose lipid in Ustilago maydis (Teichmann et al., 2007) and Pseudozyma flocculosa (Teichmann et al., 2011b). Cyp1 and Cyp2—cytochrome P450 monooxygenases, Ugt1 and Fgt1—UDP-glucose-dependent glycosyltransferases, Uat2, Fat2, and Fat3—acetyltransferases, Uat1 and Fat1—acyltransferases, Adh1—α-hydroxylase.

hydroxylase of *Ps. flocculosa*, which hydroxylate fatty acids C6 and C8 at the β position before they are transferred to the cellobiose residue (Teichmann et al., 2011b). At the last stage of ustilagic acid biosynthesis, α-hydroxylase Ahd1 catalyzes hydroxylation of the long-chain fatty acid residue at the α-position (Teichmann et al., 2011b). Flocculosin is hydroxylated at the β position. The enzyme catalyzing this hydroxylation is unknown. Neither Fhd1 nor Ahd1 fatty acid hydroxylase was responsible for this reaction (Teichmann et al., 2011b).

The expression of all cluster genes depends on Rua1, the nuclear protein of the C_2H_2 zinc finger family. The gene *rua1* is located in the same gene cluster of *U. maydis* (Teichmann et al., 2010). The similar regulatory gene *rfl1* is in the cluster of *Ps. flocculosa* (Teichmann et al., 2011a).

5.1.3 Biosynthesis of Sophorolipids

The biosynthetic pathway (Figure 5.4) of sophorolipids is well described in the reviews (Van Bogaert et al., 2007, 2011). Sophorolipid production is optimal if the medium contains simultaneously hydrophilic and hydrophobic carbon sources, e.g., glucose and fatty acids. The biosynthesis begins with the hydroxylation of fatty acids. If the medium contains triglycerides, the extracellular lipase hydrolyzes them to fatty acids, which are consumed by the cells. *St. bombicola* can also grow on *n*-alkanes and oxidize them to fatty acids. If the medium contains no fatty acids, they are formed by the known biosynthetic pathway with the involvement of acetyl-CoA. Cerulenin, the inhibitor of fatty acid synthesis *de novo*, suppressed sophorolipid biosynthesis under these conditions (Van Bogaert et al., 2008a–c).

Fatty acids must be hydroxylated to participate in sophorolipid biosynthesis. The terminal hydroxylation of fatty acids in yeasts is performed by the cytochrome P450 monooxygenase from the CYP52 family. The specific enzyme of this family, CYP52M1, which is induced under conditions of sophorolipid production, has been identified (Van Bogaert et al., 2009a,b). This enzyme is a microsomal monooxygenase forming a short electron-transport chain together with NADPH cytochrome P450 reductase (Van Bogaert et al., 2007).

Sophorolipids have fatty acid residues with 16 or 18 carbon atoms. These modifications, just as the appearance of a double bond in the fatty acid residue, do not need the enzymes specific for the pathway of sophorolipid biosynthesis but are performed by the common enzymes of fatty acid biosynthesis.

After the hydroxylation of fatty acids, two molecules of UDP-glucose are involved in the reaction. The glucose consumed from the culture medium is not included directly in the sophorolipid molecule; the glucose for sophorolipid synthesis is formed during gluconeogenesis (Hommel et al., 1994). This fact explains the inability to modify the

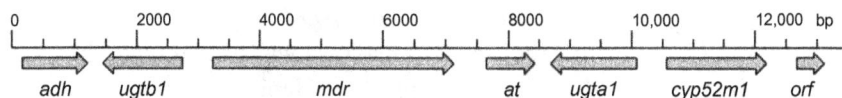

Figure 5.4 Sophorolipid biosynthesis gene cluster (Van Bogaert et al., 2013). The genes and their functions are indicated in Table 5.1 (By permission of John Wiley and Sons).

composition of the carbohydrate residue of sophorolipids by adding other sugars to the culture medium, as well as successful sophorolipid synthesis in the absence of glucose.

Two glucose transferases have been identified: Ugta1 (Saerens et al., 2011a) and Ugtb1 (Saerens et al., 2011b). The products of these genes seem to sequentially transfer glucose residues. As a result, an unacetylated sophorolipid molecule is formed. Further modifications are performed by acetyltransferase. The acetyltransferase-negative strain of *St. bombicola* secreted an unacetylated sophorolipid (Saerens et al., 2011c).

The genome of *St. bombicola* was sequenced and annotated (Van Bogaert et al., 2013). The genes of the sophorolipid biosynthesis pathway were found in the gene cluster (Figure 5.5). In addition to the genes encoding the enzymes of sophorolipid biosynthesis, the cluster contained the *mdr* gene encoding a transport protein and the *orf* gene with unknown function. All these genes are intronless. The mutant *Δadh* strain (the gene encoding the putative alcohol dehydrogenase) and the *Δorf* strain displayed neither the loss of cell viability nor changes in sophorolipid production (Van Bogaert et al., 2013). No genes encoding the enzymes responsible for internal esterification or lactonization were detected in this cluster. However, lactonization could be spontaneous or catalyzed by lactone esterases, which are not located in the cluster (Van Bogaert et al., 2013).

The biosyntheses of cellobiose lipids and sophorolipids have much in common. First, it is the formation of hydroxy fatty acids. Hydroxylation involves cytochrome P450 monooxygenases, one or several, depending on the number of hydroxyl groups. It is followed by sequential transfer of glucose residues to the fatty acid residue by UDP-glucose-dependent glycosyltransferases and, at the final step, attachment of O-substituents and acetate groups (by acetyltransferases). Because *St. bombicola* is characterized by high glycolipid production, it was used to obtain a cellobiose lipid producing strain (Roelants et al., 2013). The Ugtb1-encoding sequence in the *St. bombicola* genome was replaced by the codon-optimized analog of the *U. maydis Ugt1* gene. The strain produced 17-*O*-6′6″-diacetyl-cellobiosyl-octadecanoic acid. Thus, the acetyltransferase was able to acetylate not only sophorose but also cellobiose residue. Hence, *St. bombicola* was proposed as a platform organism for the production of new biosurfactants (Roelants et al., 2013).

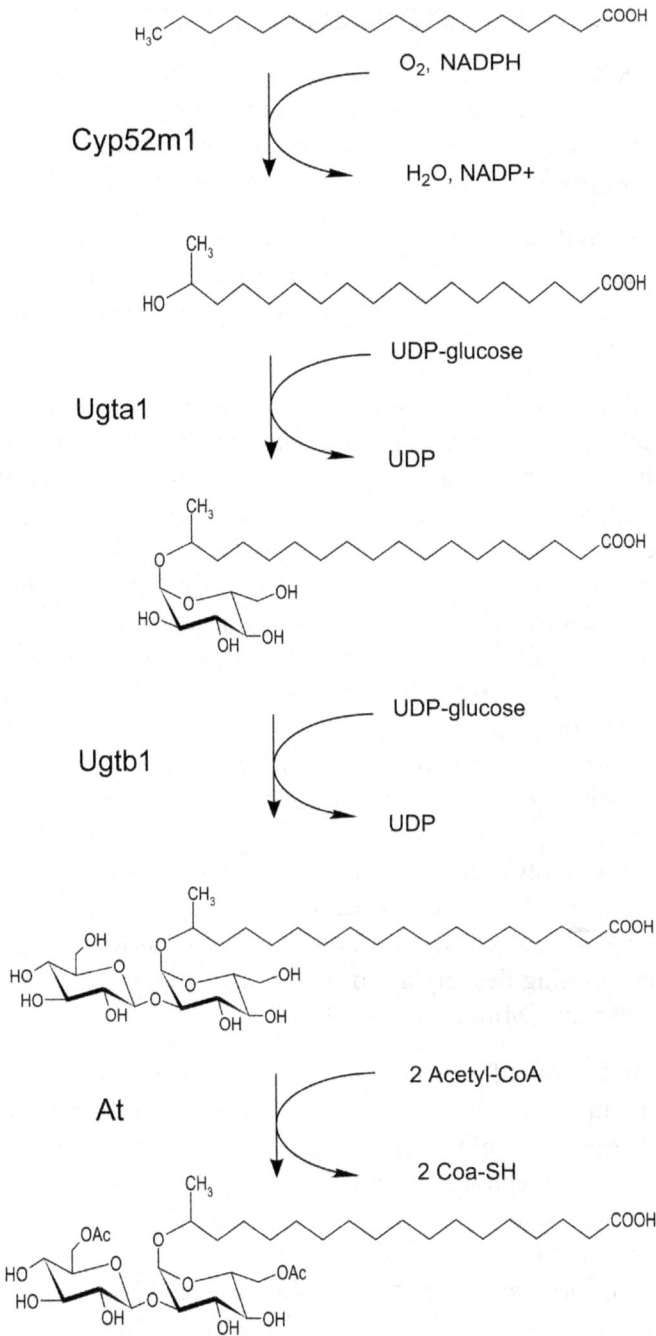

Figure 5.5 Sophorolipid biosynthetic pathway (Van Bogaert et al., 2011). Cyp52m1—cytochrome P450 mono-oxygenase, Ugta1 and Ugtb1—UDP-glucose-dependent glycosyltransferases, At—acetyltransferase.

5.2 CATABOLISM OF EXTRACELLULAR GLYCOLIPIDS

The catabolic pathways of extracellular glycolipids explain the multi-functional role of these compounds: antimicrobial compounds, biosurfactants for consumption of hydrophobic substrates, and renewable carbon reserves for survival under carbon deficiency.

Sophorolipids are metabolized by *St. bombicola* upon carbon starvation (Van Bogaert et al., 2009b). The level of sophorolipid catabolism by *St. bombicola* reached 60% in 8 days of the cultivation (Hirata et al., 2009).

The catabolism of cellobiose lipid flocculosin by producer cells has been studied (Mimee et al., 2009b). The conidia of *Ps. flocculosa* were placed into a medium lacking a carbon source but containing yeast extract, peptone, and cellobiose lipid (1 g/l); the composition of metabolic products was assayed by mass spectrometry. Flocculosin almost completely disappeared from the medium during 24 h, and its degradation products were formed: first partially-deacetylated and then fully-deacetylated cellobiose lipid. Later on, there appeared a glycolipid with glucose as a carbohydrate component and a free fatty acid. During this process, the pH value increased from 6.0 to 8.6. Alkalization results from the release of ammonium ions during deamination of amino acids, which intensifies under carbon deficiency. If pH was maintained at the level of 6.0, flocculosin remained stable. Flocculosin was also stable in alkaline buffer solutions in the absence of producer cells. It has been supposed that flocculosin degradation is provided by several enzymes secreted into the medium and the first enzyme performing deacetylation of the cellobiose residue has an alkaline pH optimum (Mimee et al., 2009b).

The catabolism of extracellular glycolipids can be performed not only by the producers but also by the fungi not synthesizing such compounds. *Fusarium moniliforme* secretes a glycosidase, which breaks the bond between the cellobiose and fatty acid residues and forms acetylated cellobiose and hydroxylated fatty acid (Eveleigh et al., 1964). Such ability may be a cause of insensitivity of the producers and other fungi to fungicidal activities of cellobiose lipids. The fungus *Hypocrea jecorina* degrades sophorolipids with the release of sophorose (Lo and Ju, 2009). It is probable that other microorganisms secreting the enzymes that cleave glycosidic bonds are able to degrade glycolipids.

CHAPTER 6

Prospects of Practical Application of Sophorolipids, Cellobiose Lipids, and MELs

6.1 APPLICATION AS MEMBRANOTROPIC AGENTS

The yeast extracellular glycolipids attract attention as a tool for target action on cellular membranes. The promising use of MELs is the improvement of gene transfection (Inoh et al., 2001, 2004). It has been demonstrated that the complex of the target DNA and the liposomes containing MEL-A and the cationic cholesterol derivative cholesterol-3-β-carboxyamidoethylene-*N*-hydroxyethylamine enables efficient penetration of the target plasmid into human cells of several cultivated lines. Compared to the commercially-available cationic liposomes, the MEL-containing liposomes provided a 50- to 70-fold increase in transfection efficiency (Inoh et al., 2001, 2004). The scheme of such experiment is presented in Figure 6.1 (Kitamoto et al., 2002). MEL-A and protamine synergistically accelerate the nuclear delivery of foreign genes and consequently promote gene transfection efficiency (Inoh et al., 2013).

We used cellobiose lipids as a membrane-damaging agent in the studies of metabolism of inorganic polyphosphates, the important regulatory biopolymers (Kulaev et al., 2004). The effect of cellobiose lipids on accumulation of these polymers was investigated to clarify the role of cellular membranes in polyphosphate biosynthesis in yeast (Kulakovskaya et al., 2008; Trilisenko et al., 2012). It was shown that some part of inorganic polyphosphates of *S. cerevisiae* cells can be synthesized only in the case of the cytoplasmic membrane integrity, while the other part consisting of shorter chain polymers is synthesized independently on plasma membrane damage.

6.2 PROSPECTS OF APPLICATION OF YEAST EXTRACELLULAR GLYCOLIPIDS IN INDUSTRY, AGRICULTURE, AND MEDICINE

A number of reviews cover the topic of practical application of biosurfactants (microbial surfactants), including extracellular yeast glycolipids (Kitamoto et al., 2002; Cameotra and Makkar, 2004; Mulligan,

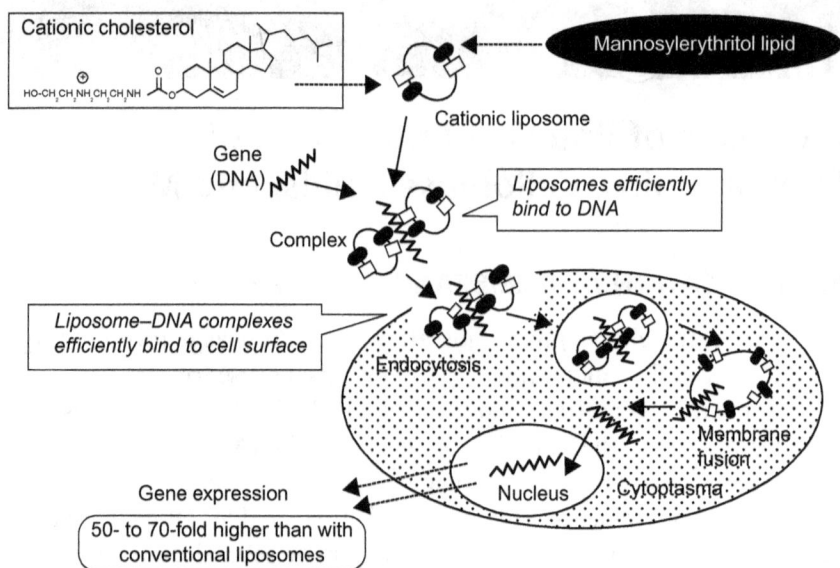

Figure 6.1 Scheme of gene delivery into mammalian cells using cationic liposomes including mannosylerythritol lipids. Kitamoto et al. (2002) by permission of Elsevier.

2005; Rodrigues et al., 2006; Muthusamy et al., 2008; Banat et al., 2010; Arutchelvi and Doble, 2011; Marchant and Banat, 2012; Cortés-Sánchez et al., 2013).

Along with bacterial surfactants, yeast surfactants attract attention due to the following properties:

— amphiphilicity
— low toxicity
— various biological activities
— biodegradability
— possibility of relatively low-cost microbiological production using nonpathogenic species and inexpensive nutrient media based on various wastes, from food industry waste to biodiesel fuel.

Different promising applications of sophorolipids and MELs associated primarily with their surfactant properties are discussed in the literature. The possible use of biosurfactants for improvement of crude oil mobility is discussed (Marchant and Banat, 2012). Glycolipids seem to be promising as emulsifying agents of low-soluble compounds (e.g., phenanthrene) in biodegradation systems and biosensors for detection and quantification of the latter (Cameotra and Makkar, 2004). The effect of sophorolipids on microbial biodegradation of low-soluble compounds has been estimated (Schippers et al., 2000). The bacteria capable of phenanthrene degradation were grown in a medium containing this compound. On addition of sophorolipids, phenanthrene biodegradation increased and no toxic effect on bacterial cells was observed. Such effect was due to enhanced phenanthrene solubility and availability for bacterial cells.

The possibilities of using glycolipids for environmental bioremediation, for example, from oil pollutions, are also considered (Cameotra and Makkar, 2004; Mulligan, 2005). The complex of glycolipids produced by *Ps. antarctica* during the growth on *n*-undecane was proposed to be used for improvement of hydrocarbon uptake by microorganisms (Hua et al., 2003). Glycolipid biosurfactants were proposed to be used for soil bioremediation from hydrophobic pollutants such as polychlorinated and polyaromatic hydrocarbons (Golyshin et al., 1999; Noordman et al., 2000). Glycolipids facilitate the release of compounds adsorbed in soil and make them available for biodegradation. Sophorolipids improve the removal of heavy metals from polluted soils (Sandrin et al., 2000; Mulligan et al., 2001), as well as inhibit the growth of blue-green algae and reduce their sorption on surfaces (Sun et al., 2004).

The potential of *St. bombicola* in treating high-fat- and oil-containing dairy industry wastewater was studied (Shah et al., 2007; Daverey and Pakshirajan, 2011). Results from the batch-operated fermentor revealed complete utilization of fats present in the wastewater within 96 h with more than 93% chemical oxygen demand (COD) removal efficiency (Daverey and Pakshirajan, 2011).

It has been proposed to use MELs to prevent the agglomeration of ice particles in refrigeration systems (Kitamoto et al., 2001b, 2002). Aggregation of ice particles results in clogging of tubes and overloading refrigerators and other freezing devices widely used in different branches of industry and in households. Biosurfactants can be adsorbed at the surface of ice particles, thereby providing their separation and stabilization and enhancing the efficiency of cooling systems. MEL produced by *Ps. antarctica* proved to be an effective reagent at a concentration of 10 mg/l. The effect of this glycolipid on ice particles is shown in Figure 6.2 (Kitamoto et al., 2001b, 2002). The authors

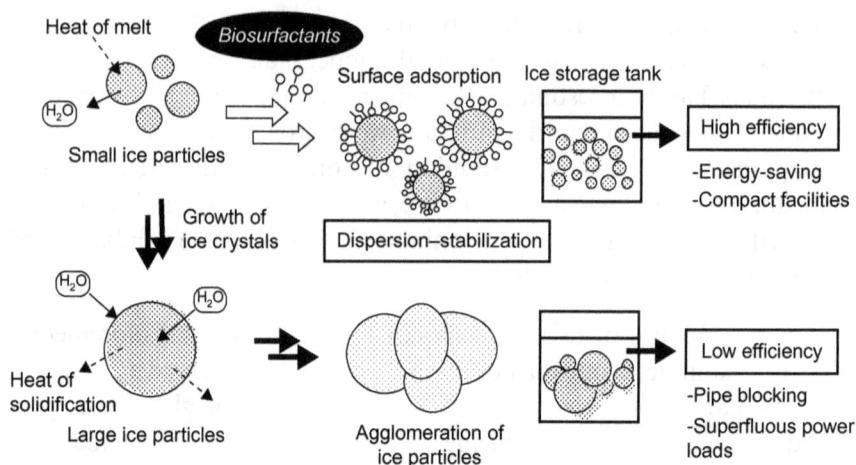

Figure 6.2 Scheme of antiagglomeration of ice particles in ice slurry system by biosurfactant. Kitamoto et al. (2002) by permission of Elsevier.

believe that this ecologically-safe reagent would be useful for refrigeration industry.

Cellobiose lipids are not yet considered as potential commercially significant surfactants, because their production by yeast is at least an order lower even in the most optimized cultivation variants, compared to sophorolipids and MELs. However, the cellobiose lipid of *Cr. humicola* has been recently obtained in an amount of 13.1 g/l (Morita et al., 2011a) and proposed to be used as a gel-forming component in production of cosmetics and in other branches of industry (Imura et al., 2012). This glycolipid was shown to form gels in different solvents, including ethanol and 1,3-butane diole, as well as in their mixtures with water. These gels are formed under mild conditions and at a temperature below 100°C, which is important for application in the production of creams, ointments, and other gel substances.

Cellobiose lipids show relatively high antifungal activity at acidic pH values. The broad spectrum of antifungal activity, including cryptococcosis and candidiasis pathogens and phytopathogenic fungi, the thermal and storage stability of cellobiose lipids make them promising compounds for the development of new fungicides. There is a patent for the so-called flocculosin, the cellobiose lipid of *Ps. flocculosa*, as an agent against yeasts and fungi pathogenic for plants, animals, and man (Belanger et al., 2004). It would certainly be interesting to use cellobiose lipids in agriculture for controlling plant diseases caused by phytopathogenic fungi such as white rot and powdery mildew, as well as for protection of the harvested fruit and vegetables from rotting. In this case, it may be prospective to use not relatively expensive cellobiose lipid preparations but colonization of plants by producer cultures. Colonization of tobacco and potato plants by *Ps. fusiformata* VKM Y-2821 resulted in enhanced resistance to the white rot pathogen *Sclerotinia sclerotiorum* (Georgievskaya et al., 2006). The strains *Ps. fusiformata* were tested against *Monilinia laxa*, the fungus causing the damage of harvested peaches. It has been shown that spraying with culture suspensions improves fruit conservation (Zhanga et al., 2010).

The basic problem of using cellobiose lipids as human and animal drugs is the absence of antifungal activity at neutral pH values (Golubev and Schabalin, 1994; Golubev et al., 2001; Puchkov et al., 2002). However, they may be rather interesting in treatment of dermatomycosis and as a component of cosmetic creams and shampoos.

The cost of glycolipid production depends both on the productivity of strains and on the price of nutrient media. The fungal extracellular glycolipids currently used in practice are sophorolipids, due to development of the relatively cheap methods of their microbiological production with a target yield of $100-400 \, g/l$ (Van Bogaert et al., 2007). The cost of the product is not a special impediment to application in cosmetology. Therefore, sophorolipids have already been extensively used in this field. The patents on the application of sophorolipids in cosmetology appeared in the late 1990s. It was proposed to use water−oil emulsions containing $0.01-30\%$ sophorolipids in cosmetics and dermatological preparations, because these compositions suppressed free radical formation, inhibited elastase activity, and possessed anti-inflammatory activity (Hillion et al., 1998). In one of these patents, it is indicated that sophorolipids are able to stimulate macrophages. The cytotoxic effect (decrease in survival by more than 90%) on fibroblasts takes place at the concentrations of $10^{-4} \, M$ and $5 \times 10^{-3} \, M$ for the lactone and acidic forms of sophorolipid, respectively. The macrophage activity is stimulated already at $10^{-6} \, M$. It was proposed to use sophorolipids as agents intensifying the fibrinolysis and desquamation of epithelium (i.e., the better peeling of dead epithelial cells) and as macrophage activators and depigmentary agents in cosmetology (Maingault, 1999).

The review by Shete et al. (2006) presents the analysis of patent literature up to 2006 and gives references to 59 patents on sophorolipids. These patents are concerned mainly with different methods of sophorolipid production, the properties of these compounds, and their application in cosmetology. Some patents suggest other promising fields of application for sophorolipids, MEL, and cellobiose lipids, which are associated with their biological activities.

One of the patents proposes to use sophorolipids as agents stimulating the growth of fibroblasts during *in vitro* cultivation (Borzeix, 2000). There is also a patent on application of sophorolipids in microbiological production of some enzymes (Gross et al., 2008). It has been shown that amylase production by *B. subtilis* increases by 39% and laccase production by *Pleurotus ostreatus* increases by 4.5-fold during their cultivation in the presence of 1 mM sophorolipids.

The application of sophorolipids for sepsis treatment was proposed (Gross, 2007). The patent describes the experiments in mice with induced sepsis and suggests intraperitoneal injections of 0.01−0.1 mg per individual, which ensure the healing of laboratory animals. The method of testing in a cell culture infected with the herpes virus showed antiviral activity characteristic of the natural sophorolipid mixture; however, the sophorolipid ethyl ester with the C-18 fatty acid residue and without any acetate groups in the sophorose residue proved to be the best one. Its semiinhibitory concentration was 0.03 μm, whereas for the natural sophorolipid mixture it was about 50 μM (Gross and Shah, 2007). In a patent entitled "Treatment and Prophylaxis of Cancer" (Gross and Bluth, 2009), it is indicated that the treatment of human pancreas carcinoma cells with increasing concentrations (from 0.5 to 2 mg/ml) of the natural mixture of *St. bombicola* sophorolipids and their selected derivatives (ethyl ester, monoacetate ethyl ester, methyl ester, the acidic form of sophorolipid, and the lactone of diacetylated sophorolipid) resulted in the necrosis and apoptosis of carcinoma cells within 24 h. The sophorolipid-conjugated gellan gum reduced gold nanoparticles showed killing effect in the glioma cell lines (Dhar et al., 2011).

For MEL, there are much fewer patents concerned with these glycolipids. One can mention the patent entitled "Biosurfactant Activators: Mannosyl Erythritol Lipid, and Production Methods" (Suzuki et al., 2010). The authors have patented the method for

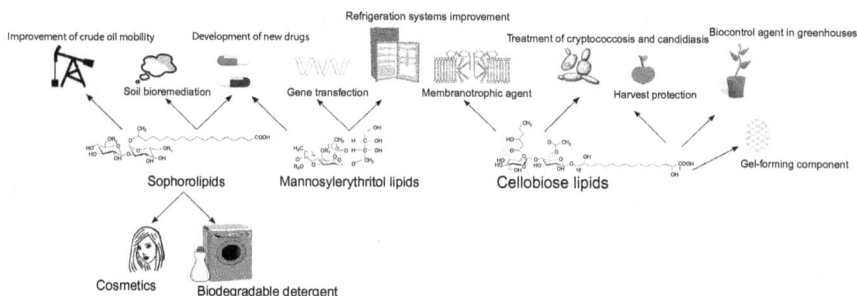

Figure 6.3 Application of sophorolipids, cellobiose lipids, and MELs: commercial products today (bottom) and prospects (top).

microbiological production of MEL and proposed to use it as an inhibitor of cell aging and a component of drugs, food additives, and cosmetics.

It must be admitted that all of the above prospective areas of sophorolipid and MEL application, including those associated with anticancer activity (Arutchelvi et al., 2008), are still far from commercialization and need further studies. It is probable that the anticancer, antiviral, and antibacterial activities of sophorolipids are not specific but may be related to their membrane-damaging and solubilizing effects.

The study of compounds responsible for the antifungal activity of yeasts is promising both with a view to gain a better understanding of the role of yeasts in the natural communities, and from a practical standpoint in the search for new agents against pathogenic yeasts and mycelial fungi.

Glycolipids secreted by yeast and fungi, due to their detergent properties and various biological activities, are promising innovative products of biotechnology for environmental protection, the food industry, agriculture, medicine, cosmetology, and other fields of human activity (Figure 6.3).

6.3 COMMERCIAL PRODUCTS BASED ON YEAST EXTRACELLULAR GLYCOLIPIDS

The product consisting of sophorolipids is mentioned in the catalogue of cosmetic components used in EC as Candida Bombicola/Glucose/Methyl Rapeseedate Ferment. It is indicated that the product is obtained by fermentation of glucose and rapeseed oil methyl esters by *C. bombicola*. It is also indicated that it has antimicrobial, antiseborrheic, deodorant, and skin-protective properties (http://ec.europa.eu/consumers/cosmetics/cosing/index.cfm?fuseaction = search.details&id = 55061). Company websites provide no information about the method of production of this substance, probably in order for this information to be a trade secret. However, it is evident from the product name that it is obtained by the cultivation of *C. bombicola* on glucose and rapeseed oil.

There is also a 40% sophorolipid solution at the market called Sopholiance (http://www.specialchem4cosmetics.com/tds/sopholiance/soliance/6580/index.aspx): "Sophorose-lipids. Shows specific anti-bacterial activity and anti-lipase activity. Used for problem skins and deodorants. Sopholiance S is recommended from 1% to 2% in a wide range of cosmetic products including oily skin products, face cleansers, makeup removers, acne prone skin care, specific care for asian skin and deodorants. This product is intended for cosmetic manufacturing and recommended for addition into ready cosmetic emulsions at a temperature below 40°C at a concentration of 0.5−3%. " The online advertisement asserts that "it is a bioactive agent for antibacterial skin protection from pathogenic microorganisms and their waste products, which maintains the normal skin flora, suppresses the propagation of *Propionibacterium acnes* causing the appearance of pustular elements, has antimicrobial action, is used for skin purification and protection from unpleasant odor caused by bacterial growth, considerably reduces sebum production, diminishes pustules and papules (after a course of treatment of no less than 28 days), and has a generally sebo-regulating effect".

The substance is recommended to be used in cosmetics for acne treatment, deodorants and antiperspirants, face cleaners (including those for makeup removal), hair drugs for treating seborrheic dermatitis, and body creams. It is mentioned that this product is used in the cosmetic agents of Korres, Bioderma, Germanie de Capuccini, Melvita, Naturopathica, Cattier and is certified by Ecocert.

The sophorolipid-based detergents are positioned by the companies as biodegradable, ecologically safe, and containing no synthetic detergents. The Ecover Company (http://www.ecover.com/) uses sophorolipids as detergents in most of the laundry and dishwasher cleaning agents, which are denoted in the catalogues as ecologically pure products, produced microbiologically at low temperatures. The Japanese company Saraya Co., Ltd. (2-2-8 Yuzato, Higashisumiyoshi-ku, Osaka, Japan) also produces a sophorolipid-containing detergent: Yashinomi Washing powder NEO. The NEO washing powder was developed on the basis of natural detergents (sophorolipids obtained by yeast fermentation of palm oil). Hence, the powder has a high washing capacity, is easily washed away and effectively biodegraded in the environment. It contains no synthetic surfactants and optical brighteners. It is appropriate for washing children's clothes. It causes no allergy. The improved sophorolipid formula of the NEO powder has the higher detergency power, compared to synthetic surfactants. Composition: detergent (sophorolipid), sodium carbonate, sodium citrate, sodium hydrocarbonate. The advantage of this washing powder is the absence of hardly degradable synthetic surfactants and phosphates, which contaminate natural waters. The main disadvantage is its relatively high price. The need of solving the problem of environmental pollution by synthetic detergents, which are low-degradable by microorganisms, and reducing the cost of sophorolipid production in future make such cleansing agents promising.

In general, yeast extracellular glycolipids are promising microbial products for industry, agriculture, and medicine. Their application can be expanded due to development of inexpensive methods of microbiological production using nutrient media on the basis of the food industry, agricultural, and biodiesel production wastes.

APPENDIX

Selected Techniques of Purification and Assay of Extracellular Yeast Glycolipids

A.1 METHODS FOR CULTIVATING PRODUCERS AND OBTAINING GLYCOLIPIDS

A.1.1 Cellobiose Lipids of Various Yeast Strains (Kulakovskaya et al., 2004, 2005, 2009)

Cellobiose lipids were produced using the yeast *Symp. paphiopedili* (Sugiyama) VKM Y-2817, *Cr. humicola* (Daszewska) Golubev VKM Y-2238, Y-1613, 9-6 (obtained by W.I. Golubev by selecting the most active cultures), X-397 and X-297 (isolated from plants of the Kedrovaya Pad State Nature Reserve in the Far East), *Ps. fusiformata* (Buhagiar) Boekhout VKM Y-2821, Y-2898, Y-2909, Ll-16, Ll-41, Ll-71, PTZ-351, and PTZ-356 (isolated from vegetation of the Prioksko-Terrasny Nature Reserve (Golubev and Golubeva, 2004)), and *Ps. graminicola* VKM Y-2938 and Ll-46 (isolated by W.I. Golubev from grasses of the Moscow region).

The cultures were maintained on wort-agar slants, stored at $0-4°C$, and periodically passed into a fresh medium. The yeast of the genus *Pseudozyma* and *Cr. humicola* were cultivated in a liquid medium containing (g/l): glucose, 10.0; $(NH_4)_2SO_4$, 1.0; yeast extract, 0.5; $MgSO_4 \cdot 7H_2O$, 0.05; $Na_2HPO_4 \cdot 12H_2O$, 10.9; pH was adjusted to 4.0 by adding citric acid. Cultivation was performed without shaking at $20-24°C$: 4 weeks for *Pseudozyma* and 2 weeks for *Cr. humicola*. The cultivation medium for *Symp. paphiopedili* contained (g/l): glucose, 10.0; $(NH_4)_2SO_4$, 1.0; yeast extract, 0.5; $MgSO_4 \cdot 7H_2O$, 0.05; succinic acid, 7.8; pH was adjusted to 4.0 by adding NaOH. Cultivation was performed without shaking at a temperature of $20-24°C$ for 4 weeks.

Antifungal compounds were obtained as follows: the culture liquid (~ 3 l) was separated from the biomass by centrifugation at $5000g$ for 40 min, filtered through GF/A fiberglass filters (Whatman), and lyophilized. Antifungal substances were extracted from the lyophilisate with methanol (400–500 ml) during 4–5 days at 4°C and undissolved components were removed by filtration through a glass filter. Methanol extracts were evaporated under vacuum at $\sim 50°C$ almost to dryness and, after addition of 250 ml of deionized water (at $\sim 4°C$), held at the same temperature for $1-3$ days for glycolipid precipitate to be formed. The precipitate was separated by filtration through a glass filter, twice washed with deionized water at $5°C$, and dissolved in methanol. If the mass spectra of the preparations showed few minor signals and their intensity was low, such preparations were used without further purification. If the mass spectrum or analytical thin-layer chromatography showed a lot of minor signals, the preparations were further purified by thin-layer chromatography on Silica gel using chloroform:methanol:water, 4:4:0.2 or 5:3:0.2. The glycolipid yield was $25-50$ mg/l.

The following method proved to be effective for *Cr. humicola*: pH of the cell suspension after cultivation was adjusted to 2.0 by HCl and kept for a few hours for precipitate formation at $0°C$, followed by 1-h centrifugation at $5000g$. The biomass precipitate with adsorbed glycolipids was washed with distilled water and again precipitated under the same conditions. The precipitate was suspended in methanol and left for 24 h at $0°C$; then the biomass was separated by filtration. The yield of cellobiose lipid was 250 mg/l. The higher yield was explained by the lower losses due to glycolipid degradation during lyophilization and evaporation.

A.1.2 Cellobiose Lipids of *Cr. humicola* (Morita et al., 2011a)

The strain *Cr. humicola* JCM 1461 was grown in shaker flasks (100 ml) with 10 ml of the medium containing 10% glucose, 0.1% peptone, 0.3% yeast extract, and 0.1% malt extract, at 200 rpm for 5 days at $25°C$.

The growth medium for cultivation in fermenters contained 10% glucose, 0.1% peptone, 0.3% yeast extract, and 0.1% wort (pH 6.0) at $25°C$. Cultivation was performed for 4 days under stirring. Then, glucose was added directly into the fermenter up to 10% concentration; the yield of cellobiose lipids after 11 days of the cultivation was 13.1 g/l (Morita et al., 2011a).

After the cultivation, glycolipids were extracted from the culture liquid with an equal volume of ethyl acetate. Then, they were purified on Silica gel (Wako-gel C-200) using column chromatography with a chloroform—acetone gradient of 10:0 to 0:10.

The yield of cellobiose lipid was 6.2 and 13.1 g/l in flask and fermenter, respectively.

A.1.3 Cellobiose Lipid Flocculosin (Mimee et al., 2009a,b)

The strain *Ps. flocculosa* DAOM 196992 was grown in the medium containing (g/l): sucrose, 35; $(NH_4)_2SO_4$, 1; KH_2PO_4, 1; $MgSO_4 \cdot 7H_2O$, 0.5; $FeSO_4 \cdot 7H_2O$, 0.01; with 50 mM citrate buffer, pH 6.0 (12.85 g/l of sodium citrate dehydrate and 1.21 g/l of citric acid). Cultivation was performed in shaker flasks (150 rpm) at 28°C. After 48-h cultivation, the culture liquid was acidified with acetic acid up to pH 2.0 and separated by filtration through Whatman paper. The precipitate was washed several times with water and washed with methanol to solubilize the glycolipid from the cells. The methanol solution was evaporated in a rotary evaporator, and the remaining syrup was washed with water and filtered through Whatman paper.

The precipitate was resuspended in water, frozen at −80°C, and lyophilized. The resultant white powder was a 99% pure flocculosin, according to mass spectrometry and thin-layer chromatography. The yield was 3 g/l.

A.1.4 Sophorolipids of *Rh. bogoriensis* (Cutler and Light, 1979)

Rh. (Torulopsis) bogoriensis was cultivated in the media containing 3% glucose, 0.15% yeast extract, and tap water. After 1-week cultivation at 25−26°C, the cells were precipitated at 14,500g for 10 min; the precipitate was suspended in a chloroform/methanol mixture (2:1) and left overnight under stirring. Then, the biomass was filtered and the organic phase was washed with water and acidified with a few drops of acetic acid to remove water-soluble components. The organic phase was separated. The maximum yield of sophorolipids was 3.6 g/l.

A.1.5 Sophorolipids of *St. bombicola* (Konishi et al., 2008)

St. bombicola NBRC 10243 and *C. batistae* CBS 8550 were grown in 300-ml flasks with 30 ml of the medium on a shaker (250 rpm) at 28°C for 3 days. The medium contained (g/l): glucose, 50; olive oil, 50; $NaNO_3$, 3; KH_2PO_4, 0.5; $MgSO_4 \cdot 7H_2O$, 0.5 yeast extract, 1 or 5; pH 6.0. After the cultivation, sophorolipids were extracted from the culture liquid with an equal volume of ethyl acetate. The organic phase was evaporated, dissolved in ethyl acetate and the glycolipid was purified by Silica gel column chromatography (Wako-gel C-200) with a chloroform−methanol gradient of 10:0 to 8:2. The yield was 6 g/l.

A.1.6 Sophorolipids of Various Yeast Strains (Kurtzman et al., 2010)

The following strains were used: *St. bombicola Y-17069*, *C. apicola Y-2481*, *C. riodocensis Y-27859*, *C. stellata Y-1449*, and *Candida* sp. NRL Y-27208. The medium contained (g/l): glucose, 100; oleic acid (100 ml/l, technical grade), 87.5; yeast extract, 1.5; NH_4Cl, 4; $KH_2PO_4 \cdot H_2O$, 1; NaCl, 0.1; $MgSO_4$, 0.5 in distilled water. The pH value was adjusted to 4.5 with KOH. Cultivation was performed at 25°C in 50-ml flasks with 10 ml of the medium at 200 rpm for 96 h or 168 h. Every day pH was acidified to 3.5 twice.

After the cultivation, the medium was acidified with 6N HCl (0.4 ml of 6N HCl was added to 10 ml of the medium) and extracted twice with 40 ml of ethyl acetate. The extract was evaporated in a rotary evaporator, resuspended in 2 ml of chloroform, and dried in a nitrogen flow. Then, unmetabolized oleic acid was extracted with hexane three times. The remainder was a fraction of sophorolipids and its purity was confirmed by MALDI-TOF mass spectrometry. The yield of sophorolipids varied depending on the strain, glucose concentration, and rate of stirring. The maximum yield for *St. bombicola* was about 70 g/l.

A.1.7 MEL (Kitamoto et al., 1990a,b; Lang, 1999)

The strain *Ps. antarctica* (CBS6821 = ATCC 32657, DSM 70725) was maintained on a potato-dextrose agar; the inoculum was grown in a medium containing (g/l): glucose, 60; NH_4NO_3, 1; KH_2PO_4, 0.5; $MgSO_4 \cdot 7H_2O$, 0.5; yeast extract, 1; tap water.

The fermentation medium contained (g/l): NH_4NO_3, 2; KH_2PO_4, 0.2; $MgSO_4 \cdot 7H_2O$, 0.2; yeast extract, 1; soybean oil, 80 ml/l; tap water.

Cultivation was performed in flasks (500 ml) in 50 ml of the medium on a shaker at 30°C for 7 days. After the cultivation, the culture medium together with the cells was acidified with 6N HCl to pH 3.0, and the cell precipitate together with the glycolipid was washed twice with water at 3000*g* for 30 min. After centrifugation and water removal, the glycolipid was extracted with methanol (300 ml per 1 l of the culture liquid); the extract was concentrated to 100 ml and twice washed with hexane to remove oil residuals. Then, 200 ml of chloroform and 100 ml of water were added, the mixture was stirred, the chloroform layer was dehydrated with sodium sulfate, and the solvent was evaporated to obtain a crude MEL preparation. Individual MELs were separated in a silica gel column equilibrated with chloroform.

A.1.8 MEL (Morita et al., 2008c)

The strain *Ps. parantarctica* JCM 11752 was used. The inoculum was grown in a medium containing 4% glucose, 0.3% $NaNO_3$, 0.03% $MgSO_4$, 0.03% KH_2PO_4, and 0.1% yeast extract (pH 6.0) at 30°C and 150 rpm for 2 days. The producing culture was grown in the same medium but with 16% soybean oil as a carbon source and cultured for 8 days under the same conditions. Sometimes, the cultivation was performed for 4 weeks in 200-ml flasks with 20 ml of the medium at 250 rpm, with extra portions of soybean oil added after each week. After the cultivation, MEL was extracted from the culture liquid with an equal volume of ethyl acetate. Individual MELs were purified by HPLC in silica gel columns.

A.2 METHODS OF ANTIFUNGAL ACTIVITY ASSAY

A.2.1 Growth Inhibition Zones on Agarized Media (Golubev and Shabalin, 1994; Kulakovskaya et al., 2003)

Antifungal activity was determined by the suppression of culture growth on glucose-peptone agar containing (g/l): glucose, 5.0; peptone, 2.5; yeast extract, 2; agar, 20; citric acid · H_2O, 6.5; Na_2HPO_4 · $12H_2O$, 13.8 (pH ~4.0). The following two methods were used.

1. "Culture-to-culture" method. The medium was preinoculated with a test culture using a glass plate spreader with 0.1 ml of liquid yeast culture. In case of mycelial fungi, the pieces of mycelium were taken from agarized medium and placed onto the surface of fresh agarized medium. Then, the producer was streaked onto the media. Petri dishes were incubated at 20−24°C for 3−5 days.
2. The level of activity of cellobiose lipid preparations was determined by applying aliquots onto sterile 5-mm GF/A discs, which then were dried and placed on the surface of agarized medium inoculated with the test cultures. The plates were incubated for 2−3 days at 20−24°C. The activity was estimated by the diameter of test culture growth inhibition zones.

A.2.2 Cell Survival Assay

The standard method for assaying cell survival is described in Kulakovskaya et al. (2005, 2009). Yeast cell suspensions ($\sim 3 \times 10^7$ cells/ml, 0.1 ml) were treated with different concentrations of glycolipids (from 0.01 to 0.8 mg/ml) in 0.5 ml of 0.04 M citrate–phosphate buffer, pH 4.0. The incubation was performed at room temperature for 30 min. The samples were inoculated on agarized medium with different dilutions depending on glycolipid concentration and test culture sensitivity. The plates were incubated at 24–30°C for 3–5 days and the colonies were counted.

A.2.3 MIC Assay

The simplest and cheapest method of antibiotic activity assay is determination of MIC. This approach is discussed in Andrews (2001). At present, the standard technique is MIC determination in immunoassay plates, i.e., the measurement of the minimum concentration of the tested compound causing growth inhibition in a liquid medium. The antibiotic, the inoculum (standardized by cell number per unit of volume), and the nutrient medium are introduced directly into the plate. It is also possible to pretreat the cells with the antibiotic before adding the nutrient medium. The method needs low amounts of nutrient media and antibiotic and yields rapid results (in 1–2 days depending on test culture growth rate). After the cultivation, optical density is determined using plate photometers or a visual test with a specific dye. The application details for this method are available at numerous sites, e.g., http://www.antimicrobialtestlaboratories.com/Minimum_Inhibitory_Concentration_Test_MIC.html.

A.3 METHODS OF MEMBRANE-DAMAGING ACTIVITY ASSAY

A.3.1 Measurement of ATP Release from Yeast Cells

ATP release from yeast cells was determined as follows (Kulakovskaya et al., 2003): various amounts of glycolipid preparations were added to the samples, each containing 0.5 ml of 0.04 M citrate-phosphate buffer, pH 4.5, and 0.05 ml of test culture cell suspension ($A_{600} = 0.6$–1), followed by incubation for 5–30 min. Then, the samples were taken and placed directly into the luminometer cuvette.

ATP was assayed by the luciferin–luciferase method. The luminometer cuvette was filled with 2 ml of the buffer (25 mM Tris–HCl,

5 mM $MgSO_4$, 0.5 mM EDTA, 0.5 mM dithiotreitol, pH 7.8), 0.05 ml of luciferin−luciferase, and 0.05 ml of the sample. The luminometer readings were recorded immediately after sample introduction. The effects of the respective amounts of methanol on ATP release were tested as well. ATP solution of the known concentration was used as a standard.

A.3.2 Staining with Bromocresol Purple

The fluorescent dye bromocresol purple (5'5'-dibromo-O-cresolsulfonaphthalene) was proposed for assessing the permeability of yeast cell membranes, because it does not stain the cells with undamaged membranes (Kurzweilova and Sigler, 1993). The cells of Cr. terreus VKM Y-2253 ($3-6 \times 10^7$ ml^{-1}) were incubated in the presence of glycolipids and 0.2 mM bromocresol purple in 0.04 M citrate-phosphate buffer, pH 4.5, for 60 min at 20°C (Kulakovskaya et al., 2003). After the incubation, the samples were centrifuged and washed with the same buffer. Then, the emission spectrum was measured in a Hitachi-850 microfluorimeter (Japan). The excitation wavelength was 325 nm, while the maximum fluorescence value used for assaying the amount of the dye bound to the cells was recorded at 655 nm.

A.3.3 Measurement of Acidification of the Medium

The cells of S. cerevisiae acidify the medium when utilizing glucose. This process is associated both with the function of H^+-ATPase and CO_2 release from the cells and is a good criterion of the cytoplasmic membrane integrity. Inhibition of this acidification is used to assess disturbances of the barrier function of the cytoplasmic membrane caused by various agents.

The cells of S. cerevisiae were treated with glycolipids in 0.04 M phosphate−citrate buffer (pH 4.0) for 30 min, then precipitated at 5000g (10 min), and twice washed with distilled water (Kulakovskaya et al., 2003). The change in pH value of the ambient medium during cell incubation with glucose was measured with a pH-meter in 2 ml of water at 30°C in 5 min after the addition of 100 mM glucose (Petrov et al., 1992).

The test for decrease in acidification of the medium by target cells was also used in the work (Mimee et al., 2009a,b).

A.3.4 Determination of K$^+$ Ion Release from Cells

The release of potassium ions from yeast cells is often used as a criterion of impairment of the cytoplasmic membrane integrity. This release
is measured by K$^+$ selective electrodes, with the yeast placed into a
thermostatic cell under stirring, and registered with a potentiometric
recorder (Kulakovskaya et al., 2008, 2010). When assessing the level of
K$^+$ ion release per their total content in the cells, the cells are exposed
to the action of silver ions or thermal treatment for 15 min at 70°C in
a water bath as a control. K$^+$ release determination was also used to
characterize the membrane-damaging activity in the work (Mimee
et al., 2009a,b).

REFERENCES

1. Andrews JM. Determination of minimum inhibitory concentrations. *J Antimicrob Chemother* 2001;1:5−16.

2. Arutchelvi J, Bhaduri S, Uppara P, Doble M. Mannosylerythritol lipids: a review. *J Ind Microbiol Biotechnol* 2008;35:1559−70.

3. Arutchelvi J, Doble M. Mannosylerythritol lipids: microbial production and their applications. In: Soberón-Chávez G, editor. Biosurfactants. Microbiology monographs, 20. New York/Heidelberg: Springer; 2011. p. 145−77.

4. Ashby RD, Nunez A, Solaiman DK, Foglia TA. Sophorolipid biosynthesis from a biodiesel-coproduct stream. *J Am Oil Chem Soc* 2005;82:625−30.

5. Ashby RD, Solaiman DK, Foglia TA. The use of fatty acid esters to enhance free acid sophorolipid synthesis. *Biotechnol Lett* 2006;28:253−60.

6. Ashby RD, Solaiman DK. The influence of increasing media methanol concentration on sophorolipid biosynthesis from glycerol-based feedstocks. *Biotechnol Lett* 2010;32:1429−37.

7. Avis TJ, Belanger RR. Specificity and mode of action of the antifungal fatty acid *cis*-9-heptadecanoic acid produced by *Pseudozyma flocculosa*. *Appl Environ Microbiol* 2001;67:956−60.

8. Avis TJ, Belanger RR. Mechanisms and means of detection of biocontrol activity of *Pseudozyma* yeasts against plant-pathogenic fungi. *FEMS Yeast Res* 2002;2:5−8.

9. Azim A, Shah V, Doncel GF, Peterson N, Gao W, Gross R. Amino acid conjugated sophorolipids: a new family of biologically active functionalized glycolipids. *Bioconjug Chem* 2006;17:1523−9.

10. Baccile N, Babonneau F, Jestin J, Pehau-Arnaudet G, Van Bogaert I. Unusual, pH-induced, self-assembly of sophorolipid biosurfactants. *ACS Nano* 2012;26:4763−76.

11. Banat IM, Franzetti A, Gandolfi I, Bestetti G, Martinotti MG, Fracchia L, et al. Microbial biosurfactants production, applications and future potential. *Appl Microbiol Biotechnol* 2010;87:427−44.

12. Beaven JM, Charpentier C, Rose AH. Production and tolerance of ethanol in relation to phospholipid fatty-acyl composition in *Saccharomyces cerevisiae*. *J Gen Microbiol* 1982;128:1447−55.

13. Bednarski W, Adamczak M, Tomasik J, Plaszczyk M. Application of oil refinery waste in the biosynthesis of glycolipids by yeast. *Bioresour Technol* 2004;95:15−8.

14. Belanger R, Labbe C, Cheng Y, McNally D: Flocculosin-CA, antimicrobial molecule, Canada Patent CA2492666, 2004.

15. Bergsson G, Arnfinnsson J, Steingrimsson O, Thormar H. *In vitro* killing of *Candida albicans* by fatty acids and monoglycerides. *Antimicrob Agents Chemother* 2001;45:3209−12.

16. Bhattacharjee SS, Haskins RH, Gorin PAJ. Location of acyl groups on two partly acylated glycolipids from strains of *Ustilago* (smut fungi). *Carbohydr Res* 1970;13:235−46.

17. Bluth MH, Kandil E, Mueller CM, Shah V, Lin YY, Zhang H, et al. Sophorolipids block lethal effects of septic shock in rats in a cecal ligation and puncture model of experimental sepsis. *Crit Care Med* 2006;34:188−95.

18. Bölker M, Basse CW, Schirawski J. *Ustilago maydis* secondary metabolism—from genomics to biochemistry. *Fungal Genet Biol* 2008;45:588–93.

19. Boros C, Katz B, Mitchell S, Pearce C, Swinbank K, Taylor D. Emmyguyacins A and B: unusual glycolipids from a sterile fungus species that inhibit the low-pH conformational change of hemagglutinin A during replication of influenza virus. *J Nat Prod* 2002;65:108–14.

20. Borzeix F: Sophorolipids as stimulating agent of dermal fibroblast metabolism, United States Patent 2000, US6057302.

21. Brajtburg J, Powderly WG, Kobayashi GS, Medoff G, Amphotericin B. Current understanding of mechanisms of action. *Antimicrob Agents Chemother* 1990;34:183–8.

22. Cameotra SS, Makkar RS. Recent applications of biosurfactants as biological and immunological molecules. *Cur Opin Microbiol* 2004;7:262–6.

23. Carballeira NM, Sanabria D, Parang K. Total synthesis and further scrutiny of the *in vitro* antifungal activity of 6-nonadecynoic acid. *Arch Pharm (Weinheim)* 2005;338:441–3.

24. Casas JA, Garcia-Ochoa F. Sophorolipid production by *Candida bombicola*: medium composition and culture methods. *J Biosci Bioeng* 1999;88:488–94.

25. Cases S, Smith SJ, Zheng YW, Myers HM, Lear SR, Sande E, et al. Identification of a gene encoding an acyl-CoA-diacylglycerol-acyltransferase, a key enzyme in triacylglycerol synthesis. *Proc Natl Acad Sci USA* 1998;95:13018–23.

26. Chen J, Song X, Zhang H, Qu JB, Miao JY. Production, structure elucidation and anticancer properties of sophorolipid from *Wickerhamiella domercqiae*. *Enzyme Microb Technol* 2006;39:501–6.

27. Chen J, Song X, Zhang H, Qu YB, Miao JY. Sophorolipid produced from the new yeast strain *Wickerhamiella domercqiae* induces apoptosis in H7402 human liver cancer cells. *Appl Microbiol Biotechnol* 2006;72:52–9.

28. Chen M, Dong C, Penfold J, Thomas RK, Smyth TJ, et al. Adsorption of sophorolipid biosurfactants on their own and mixed with sodium dodecyl benzene sulfonate, at the air/water interface. *Langmuir* 2011;27:8854–66.

29. Cheng Y, McNally DJ, Labbe C, Voyer N, Belzyle F, Belanger RR. Insertional mutagenesis of a fungal biocontrol agent led to a discovery of a rare cellobiose lipid with antifungal activity. *Appl Environ Microbiol* 2003;69:2595–602.

30. Clément-Mathieu G, Chain F, Marchand G, Bélanger RR. Leaf and powdery mildew colonization by glycolipid-producing *Pseudozyma* species. *Fungal Ecol* 2008;1:69–77.

31. Cooper DG, Paddock DA. *Torulopsis petrophilum* and surface activity. *Appl Environ Microbiol* 1983;46:1426–9.

32. Cortés-Sánchez AJ, Hernández-Sánchez H, Jaramillo-Flores ME. Biological activity of glycolipids produced by microorganisms: new trends and possible therapeutic alternatives. *Microbiol Res* 2013;168:22–32.

33. Cutler AJ, Light RJ. Regulation of hydroxydocosanoic acid sophoroside production in *Candida bogoriensis* by the levels of glucose and yeast extract in the growth medium. *J Biol Chem* 1979;254:1944–50.

34. Daniel HJ, Reuss M, Syldatk C. Production of sophorolipids in high concentration from deproteinized whey and rapeseed oil in a two stage fed batch process using *Candida bombicola* ATCC 22214 and *Cryptococcus curvatus* ATCC 20509. *Biotechnol Lett* 1998;20:1153–6.

35. Daniel HJ, Otto RT, Binder M, Reuss M, Syldatk C. Production of sophorolipids from whey: development of a two-stage process with *Cryptococcus curvatus* ATCC 20509 and *Candida bombicola* ATCC 22214 using deproteinized whey concentrates as substrates. *Appl Microbiol Biotechnol* 1999;51:40–5.

36. Daverey A, Pakshirajan K. Production, characterization, and properties of sophorolipids from the yeast *Candida bombicola* using a low-cost fermentative medium. *Appl Biochem Biotechnol* 2009;158:663–74.

37. Daverey A, Pakshirajan K. Kinetics of growth and enhanced sophorolipids production by *Candida bombicola* using a low-cost fermentative medium. *Appl Biochem Biotechnol* 2010;160:2090–101.

38. Daverey A, Pakshirajan K. Pretreatment of synthetic dairy wastewater using the sophorolipid-producing yeast *Candida bombicola*. *Appl Biochem Biotechnol* 2011;163:720–8.

39. Davila AM, Marchel R, Vandecasteele JP. Sophorose lipid fermentation with differentiated substrate supply for growth and production phases. *Appl Microbiol Biootechnol* 1997;47:496–501.

40. Deml G, Anke T, Oberwinkler F, Giannetti BM, Steglich W. Schizonellin A and B, new glycolipid from *Schizonella melanogramma*. *Phytochemistry* 1980;19:83–7.

41. Deshpande M, Daniels L. Evaluation of sophorolipid biosurfactant production by *Sandida bombicola* using animal fat. *Bioresours Technol* 1995;54:143–50.

42. Dhar S, Reddy EM, Prabhune A, Pokharkar V, Shiras A, Prasad BL. Cytotoxicity of sophorolipid-gellan gum-gold nanoparticle conjugates and their doxorubicin loaded derivatives towards human glioma and human glioma stem cell lines. *Nanoscale* 2011;3:575–80.

43. Eveleigh DE, Dateo GP, Reese ET. Fungal metabolism of complex glycosides: ustilagic acid. *J Biol Chem* 1964;239:839–44.

44. Fluharty AL, O'Brien JS. A mannose- and erythritol-containing glycolipid from *Ustilago maydis*. *Biochemistry* 1969;8:2627–32.

45. Fracchia L, Cavallo M, Martinotti MG, Banat IM. Biosurfactants and bioemulsifiers biomedical and related applications: present status and future potentials. In: Dhanjoo N, Ghista, editors. Biomedical science, engineering and technology. Rijeka, Croatia / Shanghai, China: InTech; 2012. p. 325–70.

46. Frautz V, Lang S, Wagner F. Formation of cellobiose lipids by growing and resting cells of *Ustilago maydis*. *Biotechnol Lett* 1986;8:757–62.

47. Fu SL, Wallner SR, Bowne WB, Hagler MD, Zenilman ME, Gross R, et al. Sophorolipids and their derivatives are lethal against human pancreatic cancer cells. *J Surg Res* 2008;148:77–82.

48. Fukuoka T, Morita T, Konishi M, Imura T, Kitamoto D. Characterization of new glycolipid biosurfactants, tri-acylated mannosylerythritol lipids, produced by *Pseudozyma* yeasts. *Biotechnol Lett* 2007;29:1111–8.

49. Fukuoka T, Morita T, Konishi M, Imura T, Kitamoto D. Characterization of new types of mannosylerythritol lipids as biosurfactants produced from soybean oil by a basidiomycetous yeast, *Pseudozyma shanxiensis*. *J Oleo Sci* 2007;56:435–42.

50. Fukuoka T, Kawamura M, Morita T, Imura T, Sakai H, Abe M, et al. A basidiomycetous yeast, *Pseudozyma crassa*, produces novel diastereomers of conventional mannosylerythritol lipids as glycolipid biosurfactants. *Carbohydr Res* 2008;24:2947–55.

51. Fukuoka T, Morita T, Konishi M, Imura T, Kitamoto D. A basidiomycetous yeast, *Pseudozyma tsukubaensis*, efficiently produces a novel glycolipid biosurfactant. The identification of a new diastereomer of mannosylerythritol lipid-B. *Carbohydr Res* 2008;25:555–60.

52. Fukuoka T, Yanagihara T, Imura T, Morita T, Sakai H, Abe M, et al. The diastereomers of mannosylerythritol lipids have different interfacial properties and aqueous phase behavior, reflecting the erythritol configuration. *Carbohydr Res* 2012;351:81–6.

53. Gao R, Falkeborg M, Xu X, Guo Z. Production of sophorolipids with enhanced volumetric productivity by means of high cell density fermentation. *Appl Microbiol Biotechnol* 2013;97:1103–11.

54. Georgievskaya EB, Zakharchenko NS, Bur'yanov YI, Golubev WI. The influence of yeast *Pseudozyma fusiformata* on the resistance of plants to *Sclerotinia sclerotiorum*. *Mycol Phytopatol (in Russian)* 2006;40:53–8.

55. Golubev WI. Antagonistic interactions among yeast. *Biodiversity and ecophysiology of the yeasts.* Berlin Verlag; 2006. pp. 197–219

56. Golubev W, Shabalin Y. Microcin production in *Cryptococcus humicola*. *FEMS Microbiol Lett* 1994;119:105–10.

57. Golubev WI, Golubeva EW. Yeasts of the steppe and forest communities of Prioksko-Terraced Biosphere Reserve. *Mycol Phytopatol (in Russian)* 2004;6:20–7.

58. Golubev WI, Kulakovskaya TV, Golubeva EW. The yeast *Pseudozyma fusiformata* VKM Y-2821 producing an antifungal glycolipid. *Microbiology* 2001;70:553–6.

59. Golubev WI, Kulakovskaya TV, Kulakovskaya EV, Golubev NW. The fungicidal activity of an extracellular glycolipid from *Sympodiomycopsis paphiopedili* Sugiyama et al. *Microbiology* 2004;73:724–8.

60. Golubev WI, Shcherbinina NR, Tomashevskaya MA, Golubeva EW. Antifungal activity and distribution of *Trichosporon porosum*. *Mycol Phytopatol (in Russian)* 2008;3:257–63.

61. Golubev WI, Kulakovskaya TV, Shashkov AS, Kulakovskaya EV, Golubev NW. Antifungal cellobiose lipid secreted by the epiphytic yeast *Pseudozyma graminicola*. *Microbiology* 2008;77:171–5.

62. Golyshin PM, Fredrickson HL, Giuliano L, Rothmel R, Timmis KN, Yakimov MM. Effect of novel biosurfactants on biodegradation of polychlorinated biphenyls by pure and mixed bacterial cultures. *New Microbiol* 1999;22:257–67.

63. Gorin PAJ, Spencer JFT, Tulloch AP. Hydroxy fatty acid glycosides of sophorose from *Torulopsis magnolia*. *Can J Chem* 1961;39:846–55.

64. Graner G, Hamberg M, Meijer J. Screening of oxylipins for control of oilseed rape (*Brassica napus*) fungal pathogens. *Phytochemistry* 2003;63:89–95.

65. Gross RA. Treatment of sepsis and septic shock, United States Patent 2007, US007262178B2.

66. Gross RA, Shah V. Antifungal properties of various forms of sophorolipids, United States Patent 2005, 20050164955.

67. Gross RA, Shah V. Anti-herpes virus properties of various form of sophorolipids, World Intellectual Property Organization 2007, WO2007/ 130738 A1.

68. Gross RA, Shah V, Nerud F, Madamwars D. Sophorolipids as protein inducers and inhibitors in fermentation medium, United States Patent Application 2008, 20080076165.

69. Gross RA, Bluth MH. Treatment and prophylaxix of cancer.,United States Patent 2009, US 2009/0186835 A1.

70. Günter M, Zibek S, Hirth T, Rupp S. Synthese and Optimierung von Cellobioselipiden and Mannosylerythrytollipiden. *Chem Ingen Tech* 2010;82:1215–21.

71. Gupta R, Prabhune AA. Structural determination and chemical esterification of the sophorolipids produced by *Candida bombicola* grown on glucose and α-linolenic acid. *Biotechnol Lett* 2012;34:701–7.

72. Hammami W, Labbé C, Chain F, Mimee B, Bélanger RR. Nutritional regulation and kinetics of *flocculosin synthesis by Pseudozyma flocculosa*. *Appl Microbiol Biotechnol* 2008;80:307–15.

73. Hammami W, Chain F, Michaud D, Bélanger RR. Proteomic analysis of the metabolic adaptation of the biocontrol agent *Pseudozyma flocculosa* leading to glycolipid production. *Proteome Sci* 2010;8:3–9.

74. Hammami W, Castro CQ, Rémus-Borel W, Labbé C, Bélanger RR. Ecological basis of the interaction between *Pseudozyma flocculosa* and powdery mildew fungi. *Appl Environ Microbiol* 2011;77:926–33.

75. Hardin R, Pierre J, Schulze R, Mueller CM, Fu SL, Wallner SR, et al. Sophorolipids improve sepsis survival: effects of dosing and derivatives. *J Surg Res* 2007;142:314–9.

76. Haskins RH. Biochemistry of the Ustilaginales: I. Preliminary cultural cultural studies of *Ustilago zeae*. *Can J Res* 1950;28:213–23.

77. Haskins RH, Thorn JA. Biochemistry of the Ustilaginales. VII. Antibiotic activity of ustilagic acid. *Can J Botany* 1951;29:585–92.

78. Haskins RH, Thorn JA, Boothroyd B. Biochemistry of Ustilaginales. XI Metabolic products of *Ustilago zeae* in submerged culture. *Can J Microbiol* 1955;1:749–56.

79. Herve A, Rousseaux I, Charpentier C. Relationship between ethanol tolerance, lipid composition and plasma membrane fluidity in *Saccharomyces cerevisiae* and *Kloeckera apiculata*. *FEMS Microbiol Lett* 1994;124:17–22.

80. Hewald S, Josephs K, Bolker M. Genetic analysis of biosurfactant prduction in *Ustilago maydis*. *Appl Environ Microbiol* 2005;71:3033–40.

81. Hewald S, Linne U, Scherer M, Marahiel MA, Kamper J, Bolker M. Identification of a gene cluster for biosynthesis of mannosylerythritol lipids in the basidiomycetous fungus *Ustilago maydis*. *Appl Environ Microbiol* 2006;72:5469–77.

82. Hillion G, Marchal R, Stoltz C, Borzeix F. Use of a sophorolipid to provide free radical formation inhibiting activity or elastase inhibiting activity, United States Patent 1998, US-5756471.

83. Hirata Y, Ryu M, Oda Y, Igarashi K, Nagatsuka A, Furuta T, et al. Novel characteristics of sophorolipids, yeast glycolipid biosurfactants, as biodegradable low-foaming surfactants. *J Biosci Bioeng* 2009;108:142–6.

84. Hommel R, Stuwer O, Stuber W, Haferburg DM, Kleber HP. Production of water-soluble surface active exolipids by *Torulopsis apicola*. *Appl Microbiol Biotechnol* 1987;26:199–205.

85. Hommel RK, Weber L, Weiss A, Himmelreich U, Rilke O, Kleber H-P. Production of sophorose lipid by *Candida (Torulopisis) apicola* grown on glucose. *J Biotechnol* 1994;33:147–55.

86. Hou CT, Forman RJ. Growth inhibition of plant pathogenic fungi by hydroxy fatty acids. *J Industr Microbiol Biotechnol* 2000;24:275–6.

87. Hu Y, Ju L-K. Sophorolipid production from different lipid precursors observed by LC-MS. *Enzyme Microb Technol* 2001;29:593–601.

88. Hua Z, Chen J, Lun S, Wang X. Influence of biosurfactants produced by *Candida antarctica* on surface properties of microorganism and biodegradation of *n*-alkanes. *Water Res* 2003;37:4143–50.

89. Hurley R, de Louvois J, Mulhall A. Yeast as human and animal pathogens. In: Rose AH, Harrison JS, editors. *The yeasts, vol 1: biology of yeasts*. 2nd edition. New York, NY: Academic Press; 1987. p. 207–81.

90. Ikeda Y, Sunakawa T, Tsuchiya S, Kondo M, Okamoto K. Toxicological studies on sophorolipid derivatives. (II). Acute toxicity, eye irritation, primary skin irritation, skin sensitization, phototoxicity, photosensitization, mutagenicity of polyoxypropylene (12) [(2′-0-beta-D-glucopyranosyl-beta-D-glucopyranosyl) oxy-] fatty acid ester-]. *J Toxicol Sci* 1986;11:197–211.

91. Ikeda Y, Sunakawa T, Okamoto K, Hirayama A. Toxicological studies on sophorolipid derivatives. (II). Subacute toxicity study of polyoxypropylene (12) [2′-0-beta-D-glucopyranosyl-beta-D-glucopyranosyl) oxy-] fatty acid ester-]. *J Toxicol Sci* 1986;11:213–24.

92. Im JH, Nakane T, Yanagishita H, Ikegami T, Kitamoto D. Mannosylerythritol lipid, a yeast extracellular glycolipid, shows high binding affinity towards human immunoglobulin G. *BMC Biotechnol* 2001;1:11−5.

93. Im JH, Yanagishita H, Ikegami T, Takeyama Y, Idemoto Y, Koura N, et al. Mannosylerythritol lipids, yeast glycolipid biosurfactants, are potential affinity ligand materials for human immunoglobulin G. *J Biomed Mater Res* 2003;65:379−85.

94. Imura T, Ohta N, Inoue K, Yagi N, Negishi H, Yanagishita H, et al. Naturally engineered glycolipid biosurfactants leading to distinctive self-assembled structures. *Chemistry* 2006;12:2434−40.

95. Imura T, Masuda Y, Minamikawa H, Fukuoka T, Konishi M, Morita T, et al. Enzymatic conversion of diacetylated sophoroselipid into acetylated glucoselipid: surface-active properties of novel bolaform biosurfactants. *J Oleo Sci* 2010;59:495−501.

96. Imura T, Kawamura D, Ishibashi Y, Morita T, Sato S, Fukuoka T, et al. Low molecular weight gelators based on biosurfactants, cellobiose lipids by *Cryptococcus humicola*. *J Oleo Sci* 2012;61:659−64.

97. Inoh Y, Kitamoto D, Hirashima N, Nakanishi M. Biosurfactants of MEL-A increase gene transfection mediated by cationic liposomes. *Biochem Biophys Res Commun* 2001;289:57−61.

98. Inoh Y, Kitamoto D, Hirashima N, Nakanishi M. Biosurfactant MEL-A dramatically increases gene transfection via membrane fusion. *J Control Release* 2004;94:423−31.

99. Inoh Y, Furuno T, Hirashima N, Kitamoto D, Nakanishi M. Synergistic effect of a biosurfactant and protamine on gene transfection efficiency. *Eur J Pharm Sci* 2013;49:1−9.

100. Isoda H, Kitamoto D, Shinmoto H, Matsumoto M, Nakahara T. Microbial extracellular glycolipid induction of differentiation and inhibition of protein kinase C activity of human promyelocytic leukemia cell line HL60. *Biosci Biotechnol Biochem* 1997;61:609−14.

101. Isoda H, Shinmoto H, Matsumura M, Nakahara T. The neurite initiating effect of microbial extracellular glycolipids in PC12 cells. *Cytotechnology* 1999;31:163−70.

102. Ito S, Inoue S. Sophorolipids from *Torulopsis bombicola*: possible relation to alkane uptake. *Appl Environ Microbiol* 1982;43:1278−83.

103. Ivanov A.J., Vagabov V.M., Fomchenkov V.M., Kulaev I.S. Study of the influence of polyphosphates of cell envelope on the sensitivity of yeast *Saccharomyces carlsbergensis* to the cytyl-3-methylammonium bromide. Microbiologiia 1996; **65**: 611−616.

104. Kabara JJ, Vrable R. Antimicrobial lipids: natural and synthetic fatty acids and monoglycerides. *Lipids* 1977;12:753−9.

105. Kakugawa K, Tamai M, Imamura K, Miyamoto K, Miyoshi S, Morinaga Y, et al. Isolation of yeast *Kurtzmanomyces* sp. I-11, novel producer of mannosylerythritol lipid. *Biosci Biotechnol Biochem* 2002;**66**:188−91.

106. Kämper J, Kahmann R, Bölker M, Ma LJ, Brefort T, Saville BJ, et al. Insights from the genome of the biotrophic fungal plant pathogen *Ustilago maydis*. *Nature* 2006;444:97−101.

107. Kim HS, Yoon BD, Choung DH, Oh HM, Katsuragi T, Tani Y. Characterization of a biosurfactant, mannosylerythritol lipid produced from *Candida* sp. SY16. *Appl Microbiol Biotechnol* 1999;52:713−21.

108. Kim HS, Jeon JW, Kim BH, Ahn CY, Oh HM, Yoon BD. Extracellular production of a glycolipid biosurfactant, mannosylerythritol lipid by *Candida* sp. SY16 using fed-batch fermentation. *Appl Microbiol Biotechnol* 2006;70:391−6.

109. Kim YB, Yun HS, Kim K. Enhanced sophorolipid production by feeding-rate-controlled fed-batch culture. *Bioresour Technol* 2009;100:6028−32.

110. Kitamoto D, Akiba S, Hioki C, Tabuchi T. Extracellular accumulation of mannosylerythritol lipids by a strain of *Candida antarctica*. *Agric Biol Chem* 1990;**54**:31−6.

111. Kitamoto D, Haneishi K, Nakahara T, Tabuchi T. Production of mannosylerythritol lipids by *Candida antarctica* from vegetable oils. *Agric Biol Chem* 1990;54:37−40.

112. Kitamoto D, Fujishiro K, Yanagishita H, Nakane T, Nakahara T. Production of mannosylerythritol lipids as biosurfactants by resting cells of *Candida antarctica*. *Biotechnol Lett* 1992;14:305−10.

113. Kitamoto D, Nakane T, Nakao N, Nakahara T, Tabuchi T. Intracellular accumulation of mannosylerythritol lipids as storage materials by *Candida antarctica*. *Appl Microbiol Biotechnol* 1992;36:768−72.

114. Kitamoto D, Nemoto T, Yanagishita H, Nakane T, Kitamoto HK, Nakahara T. Fatty acid metabolism of mannosylerythritol lipids as biosurfactants produced by *Candida antarctica*. *J Jpn Oil Chem Soc* 1993;42:346−58.

115. Kitamoto D, Yanagishita H, Hayara K, Kitamoto HK. Effect of cerulenin on the production of mannosylerythritol lipids as biosurfactants by *Candida antarctica*. *Biotechnol Lett* 1995;17:25−30.

116. Kitamoto D, Yanagishita H, Hayara K, Kitamoto HK. Contribution of a chain-shortening pathway to the biosynthesis of the fatty acids of mannosylerythritol lipid (biosurfactant) in the yeast *Candida antarctica:* effect of P-oxidation inhibitors on biosurfactant synthesis. *Biotechnol Lett* 1998;20:813−8.

117. Kitamoto D, Yokoshima T, Yanagishita H, Haraya K, Kitamoto HK. Formation of glycolipid biosurfactant, mannosylerythritol lipid, by *Candida antarctica* from aliphatic hydrocarbons via subterminal oxidation pathway. *J Jpn Oil Chem Soc* 1999;48:1377−84.

118. Kitamoto D, Sangita G, Ourisson G, Nakatani Y. Formation of giant vesicles from diacylmannosylerythritols and their binding to concanavalin A. *Chem Commun* 2000;10:861−2.

119. Kitamoto D, Ikegami T, Suzuki T, Sasaki A, Takeyama Y, Idemoto Y, et al. Microbial conversion of *n*-alkanes into glycolipid biosurfactants, mannosylerythritol lipids, by *Pseudozyma (Candida antarctica)*. *Biotechnol Lett* 2001;23:1709−14.

120. Kitamoto D, Yanagishita H, Endo A, Nakaiwa M, Nakane T, Akiya T. Remarkable anti-agglomeration effect of a yeast biosurfactant, diacylmannosylerythritol, on ice-water slurry for cold thermal storage. *Biotechnol Prog* 2001;17:362−5.

121. Kitamoto D, Isoda H, Nakahara T. Functions and potential applications of glycolipid biosurfactants from energy-saving materials to gene delivery carriers. *J Biosci Bioeng* 2002;94:187−201.

122. Kodedova M, Sigler K, Lemire BD, Gaskova D. Fluorescence method for determining the mechanism and speed of action of surface-active drugs on yeast cells. *Biotechniques* 2011;50:58−63.

123. Konishi M, Morita T, Fukuoka T, Imura T, Kakugawa K, Kitamoto D. Efficient production of mannosylerythritol lipids with high hydrophilicity by *Pseudozyma hubeiensis* KM-59. *Appl Microbiol Biotechnol* 2007;78:37−46.

124. Konishi M, Fukuoka T, Morita T, Imura T, Kitamoto D. Production of new types of sophorolipids by *Candida batistae*. *J Oleo Sci* 2008;57:359−69.

125. Konishi M, Nagahama T, Fukuoka T, Morita T, Imura T, Kitamoto D, et al. Yeast extract stimulates production of glycolipid biosurfactants, mannosylerythritol lipids, by *Pseudozyma hubeiensis* SY62. *J Biosci Bioeng* 2011;111:702−5.

126. Kubota S, Takeo I, Kume K, Kanai M, Shitamukai A, Mizunuma M, et al. Effect of ethanol on cell growth of budding yeast: genes that are important for cell growth in the presence of ethanol. *Biosci Biotechnol Biochem* 2004;68:968−72.

127. Kulaev IS, Vagabov VM, Kulakovskaya TV. *The biochemistry of inorganic polyphosphates.* Chichester: Wiley; 2004.

128. Kulakovskaya TV, Kulakovskaya EV, Golubev WI. ATP leakage from yeast cells treated by extracellular glycolipids of *Pseudozyma fusiformata*. *FEMS Yeast Res* 2003;3:401–4.

129. Kulakovskaya TV, Shashkov AS, Kulakovskaya EV, Golubev WI. Characterization of antifungal glycolipid secreted by the yeast *Sympodiomycopsis paphiopedili*. *FEMS Yeast Res* 2004;5:247–52.

130. Kulakovskaya TV, Shashkov AS, Kulakovskaya EV, Golubev WI. Ustilagic acid secretion by *Pseudozyma fusiformata* strains. *FEMS Yeast Res* 2005;5:919–23.

131. Kulakovskaya EV, Golubev VI, Kulaev IS. Extracellular antifungal glycolipids of *Cryptococcus humicola* yeasts. *Dokl Biol Sci* 2006;410:393–5.

132. Kulakovskaya EV, Kulakovskaya TV, Golubev WI, Shashkov AS, Grachev AA, Nifantiev NE. Fungicidal activity of cellobiose lipids from culture broth of yeast *Cryptococcus humicola* and *Pseudozyma fusiformata*. *Russian J Bioorg Chem* 2007;33:156–60.

133. Kulakovskaya TV, Shashkov AS, Kulakovskaya EV, Golubev WI, Zinin AI, et al. Structures and fungicidal activities of cellobiose lipids. In: Sasaki D, editor. *Structures and fungicidal activities of cellobiose lipids. Glycolipids: new research*. New York, NY: Nowa Publisher; 2008. p. 171–84.

134. Kulakovskaya EV, Ivanov AY, Kulakovskaya TV, Vagabov VM, Kulaev IS. Effects of cellobiose lipid B on *Saccharomyces cerevisiae* cells: K$^+$ leakage and inhibition of polyphosphate accumulation. *Microbiology* 2008;77:288–92.

135. Kulakovskaya TV, Kulakovskaya EV, Golubev WI. Antifungal activities of glycolipids secreted by yeast. In: de Cista, Bezerra P, editors. *Fungicides: chemistry, environmental impact and health effect*. New York, NY: Nova Biomedical Books; 2009. p. 13–25.

136. Kulakovskaya TV, Shashkov AS, Kulakovskaya EV, Golubev WI, Zinin AI, Tsvetkov YE, et al. Extracellular cellobiose lipid from yeast and their analogues: structures and fungicidal activities. *J Oleo Sci* 2009;58:133–40.

137. Kulakovskaya TV, Golubev WI, Tomashevskaya MA, Kulakovskaya EV, Shashkov AS, Grachev AA, et al. Production of antifungal cellobiose lipids by *Trichosporon porosum*. *Mycopathologia* 2010;169:117–24.

138. Kulakovskaya EV, Vagabov VM, AYu Ivanov, Trilisenko LV, Kulakovskaya TV, Kulaev IS. Inorganic polyphosphates and sensitivity of *Saccharomyces cerevisiae* cells to membrane-damaging agents. *Microbiology* 2011;80:10–4.

139. Kurtzman CP, Price NP, Ray KJ, Kuo TM. Production of sophorolipid biosurfactants by multiple species of the *Starmerella (Candida) bombicola* yeast clade. *FEMS Microbiol Lett* 2010;311:140–6.

140. Kurz M, Eder C, Isert D, Li Z, Paulus EF, Schiell M, et al. Ustilipids, acylated beta-D-mannopyranosyl D-erythritols from *Ustilago maydis* and *Geotrichum candidum*. *J Antibiot (Tokyo)* 2003;56:91–101.

141. Kurzweilova H, Sigler K. Fluorescent staining with bromocresol purple: a rapid method for determining yeast cell dead count developed as an assay of killer toxin activity. *Yeast* 1993;9:1207–13.

142. Laine RA, Griffin PFS, Sweeley CC, Brannan PJ. Monoglucosyloxydecenoic acid, a glycolipid from *Aspergillus niger*. *Biochemistry* 1972;11:2267–71.

143. Lang S. Production of microbial glycolipids. In: Bucke C, editor. *Methods in biotechnology. Carbohydrate biotechnology protocols*, vol 10. Totowa, NJ: Humana Press Inc.; 1999.

144. Lang S, Wagner F. Structures and properties of biosurfactants. In: Kosaric N, Cairns WL, Gray NCC, editors. *Biosurfactants and biotechnology*. New York, NY: Marcel Dekker Inc; 1987. p. 21–45.

145. Lang S, Katsiwela E, Wagner F. Antimicrobial affects of biosurfactants. *Fett Wiss Technol—Fat Sci Technol* 1989;91:363–6.

146. Langer O, Palme O, Wray V, Tokuda H, Lang S. Production and modification of bioactive biosurfactants. *Process Biochem* 2006;41:2138–45.

147. Lemieux RU. Biochemistry of the ustilaginales. III. The degradation products and proof of the chemical heterogeneity of ustilagic acid. *Can J Chem* 1951;29:415–25.

148. Lemieux RU, Thorn JA, Brice C, Haskins RH. Biochemistry of the ustilaginales. II. Isolation and partial characterization of ustilagic acid. *Can J Chem* 1951;29:409–14.

149. Li H, Ma X, Shao L, Shen J, Song X. Enhancement of sophorolipid production of *Wickerhamiella domercqiae* var. sophorolipid CGMCC 1576 by low-energy ion beam implantation. *Appl Biochem Biotechnol* 2012;167:510–23.

150. Lo CM, Ju LK. Sophorolipid–induced cellulose production in cocultures of *Hypocrea jecorina* and *Candida bombicola*. *Enzyme Microb Technol* 2009;44:107–11.

151. Ma X, Li H, Shao LJ, Shen J, Song X. Effects of nitrogen sources on production and composition of sophorolipids by *Wickerhamiella domercqiae* var. sophorolipid CGMCC 1576. *Appl Microbiol Biotechnol* 2011;91:1623–32.

152. Ma X, Li H, Song X. Surface and biological activity of sophorolipid molecules produced by *Wickerhamiella domercqiae* var. sophorolipid CGMCC 1576. *J Colloid Interface Sci* 2012;376:165–72.

153. Maingault M. Utilization of sophorolipids as therapeutically active substances or cosmetic products, in particular for the treatment of the skin, United States Patent 1999, US005981497A.

154. Marchand G, Rémus-Borel W, Chain F, Hammami W, Belzile F, Bélanger RR. Identification of genes potentially involved in the biocontrol activity of *Pseudozyma flocculosa*. *Phytopathology* 2009;99:1142–9.

155. Marchant R, Banat IM. Microbial biosurfactants: challenges and opportunities for future exploitation. *Trends Biotechnol* 2012;30:558–65.

156. Mimee B, Labbe C, Pelletier R, Belanger RR. Antifungal activity of flocculosin, a novel glycolipid isolated from *Pseudozyma flocculosa*. *Antimicrob Agents Chemother* 2005;49:1597–9.

157. Mimee B, Pelletier R, Bélanger RR. *In vitro* antibacterial activity and antifungal mode of action of flocculosin, a membrane-active cellobiose lipid. *J Appl Microbiol* 2009;107:989–96.

158. Mimee B, Labbé C, Bélanger RR. Catabolism of flocculosin, an antimicrobial metabolite produced by *Pseudozyma flocculosa*. *Glycobiology* 2009;19:995–1001.

159. Mishra P, Prasad R. Relationship between ethanol tolerance and fatty-acyl composition of *Saccharomyces cerevisiae*. *Appl Microbiol Biotechnol* 1989;30:294–8.

160. Morita T, Konishi M, Fukuoka T, Imura T, Kitamoto D. Discovery of *Pseudozyma rugulosa* NBRC 10877 as a novel producer of the glycolipid biosurfactants, mannosylerithritol lipids, based on rDNA sequence. *Appl Microbiol Biotechnol* 2006;73:305–13.

161. Morita T, Konishi M, Fukuoka T, Imura T, Kitamoto D. Analysis of expressed sequence tags from the anamorphic basidiomycetous yeast, *Pseudozyma antarctica*, which produces glycolipid biosurfactants, mannosylerythritol lipids. *Yeast* 2006;15:661–71.

162. Morita T, Konishi M, Fukuoka T, Imura T, Kitamoto HK, Kitamoto D. Characterization of the genus *Pseudozyma* by the formation of glycolipid biosurfactants, mannosylerythritol lipids. *FEMS Yeast Res* 2007;7:286–92.

163. Morita T, Konishi M, Fukuoka T, Imura T, Kitamoto D. Identification of *Ustilago cynodontis* as a new producer of glycolipid biosurfactants, mannosylerythritol lipids, based on ribosomal DNA sequences. *J Oleo Sci* 2008;57:549–56.

164. Morita T, Konishi M, Fukuoka T, Imura T, Kitamoto D. Production of glycolipid biosurfactants, mannosylerythritol lipids, by *Pseudozyma siamensis* CBS 9960 and their interfacial properties. *J Biosci Bioeng* 2008;105:493–502.

165. Morita T, Konishi M, Fukuoka T, Imura T, Sakai H, Kitamoto D. Efficient production of di- and tri-acylated mannosylerythritol lipids as glycolipid biosurfactants by *Pseudozyma parantarctica* JCM 11752(T). *J Oleo Sci* 2008;57:557–65.

166. Morita T, Konishi M, Fukuoka T, Imura T, Yamamoto S, Kitagawa M, et al. Identification of *Pseudozyma graminicola* CBS 10092 as a producer of glycolipid biosurfactants, mannosylerythritol lipids. *J Oleo Sci* 2008;57:123–31.

167. Morita T, Fukuoka T, Imura T, Kitamoto D. Production of glycolipid biosurfactants by basidiomycetous yeasts. *Biotechnol Appl Biochem* 2009;53:39–49.

168. Morita T, Fukuoka T, Konishi M, Imura T, Yamamoto S, Kitagawa M, et al. Production of a novel glycolipid biosurfactant, mannosylmannitol lipid, by *Pseudozyma parantarctica* and its interfacial properties. *Appl Microbiol Biotechnol* 2009;83:1017–25.

169. Morita T, Kitagawa M, Suzuki M, Yamamoto S, Sogabe A, Yanagidani S, et al. A yeast glycolipid biosurfactant, mannosylerythritol lipid, shows potential moisturizing activity toward cultured human skin cells: the recovery effect of MEL-A on the SDS-damaged human skin cells. *J Oleo Sci* 2009;58:639–42.

170. Morita T, Ito E, Kitamoto HK, Takegawa K, Fukuoka T, Imura T, et al. Identification of the gene PaEMT1 for biosynthesis of mannosylerythritol lipids in the basidiomycetous yeast *Pseudozyma antarctica*. *Yeast* 2010;27:905–17.

171. Morita T, Kitagawa M, Yamamoto S, Sogabe A, Imura T, Fukuoka T, et al. Glycolipid biosurfactants, mannosylerythritol lipids, repair the damaged hair. *J Oleo Sci* 2010;59:267–72.

172. Morita T, Kitagawa M, Yamamoto S, Suzuki M, Sogabe A, Imura T, et al. Activation of fibroblast and papilla cells by glycolipid biosurfactants, mannosylerythritol lipids. *J Oleo Sci* 2010;59:451–5.

173. Morita T, Ishibashi Y, Fukuoka T, Imura T, Sakai H, Abe M, et al. Production of glycolipid biosurfactants, cellobiose lipids, by *Cryptococcus humicola* JCM 1461 and their interfacial properties. *Biosci Biotechnol Biochem* 2011;75:1597–9.

174. Morita T, Ishibashi Y, Hirose N, Wada K, Takanashi M, Fukuoka T, et al. Production and characterization of a glycolipid biosurfactant, mannosylerythritol lipid B, from sugarcane juice by *Ustilago scitaminea* NBRC 32730. *Biosci Biotechnol Biochem* 2011;75:1371–6.

175. Morita T, Ogura Y, Takashima M, Hirose N, Fukuoka T, Imura T, et al. Isolation of *Pseudozyma churashimaensis* sp. nov., a novel ustilaginomycetous yeast species as a producer of glycolipid biosurfactants, mannosylerythritol lipids. *J Biosci Bioeng* 2011;112:137–44.

176. Morita Y, Tadokoro S, Sasai M, Kitamoto D, Hirashima N. Biosurfactant mannosylerythritol lipid inhibits secretion of inflammatory mediators from RBL-2H3 cells. *Biochim Biophys Acta* 2011;1810:1302–8.

177. Morita T, Fukuoka T, Imura T, Kitamoto D. Formation of the two novel glycolipid biosurfactants, mannosylribitol lipid and mannosylarabitol lipid, by *Pseudozyma parantarctica* JCM 11752(T). *Appl Microbiol Biotechnol* 2012;96:931–8.

178. Morita T, Koike H, Koyama Y, et al. Genome sequence of the basidiomycetous yeast *Pseudozyma antarctica* T-34, a producer of the glycolipid biosurfactants mannosylerythritol lipids. *Genome Announc* 2013;1:e0006413. doi:10.1128/genomeA.00064-13.

179. Mulligan CN. Environmental applications of biosurfactants. *Environ Pollut* 2005;133:183–98.

180. Mulligan CN, Raymond NY, Gibbs BF. Heavy metal removal from sediments by biosurfactants. *J Hazard Mater* 2001;85:112–25.

181. Muthusamy K, Gopalakrishnan S, Ravi TK, Sivachidambaram P. Biosurfactants: properties, commercial production and application. *Curr Sci* 2008;94:736–47.

182. Ng AW, Wasan KM, Lopez-Berestein G. Development of liposomal polyene antibiotics: an historical perspective. *J Pharm Pharm Sci* 2003;6:67–83.

183. Nguyen TTL, Edelen A, Neighbors BM, Sabatini DA. Biocompatible lecithin-based microemulsion with rhamnolipid and sophorolipid biosurfactants: formulation and potential application. *J Colloid Interface Sci* 2010;348:498–504.

184. Niwano M, Mirumachi E, Uramoto M, Isono K. Fatty acid as inhibitors of microbial cell wall synthesis. *Agric Biol Chem* 1984;48:1359–60.

185. Noordman WH, Bruining J-P, Wietzes P, Janssen DK. Facilitated transport of a PAH mixture by a rhamnolipid biosurfactant in porous silica matrices. *J Contarn Hydrol* 2000;44:119–40.

186. Nunez A, Foglia TA, Ashby R. Enzymatic synthesis of a galactopyranose sophorolipid fatty acid-ester. *Biotechnol Lett* 2003;25:1291–7.

187. Nunez A, Ashby R, Foglia TA, Solaiman DK. LC/MS analysis and lipase modification of the sophorolipids produced by *Rhodotorula bogoriensis*. *Biotechnol Lett* 2004;26:1087–93.

188. Otto RT, Daniel HJ, Pekin G, Muller-Decker K, Furstenberger G, Reuss M, et al. Production of sophorolipids from whey. II. Product composition, surface active properties, cytotoxicity and stability against hydrolases by enzymatic treatment. *Appl Microbiol Biotechnol* 1999;52:495–501.

189. Pekin G, Vardar-Sukan F, Kozaric N. Production of sophorolipids from *Candida bombicola* ATCC 22214 using Turkish corn oil and honey. *Engl Life Sci* 2005;5:357–62.

190. Permyakov S, Suzina N, Valiakhmetov A. Activation of H + -ATPase of the plasma membrane of *Saccharomyces cerevisiae* by glucose: the role of sphingolipid and lateral enzyme mobility. *PLoS One* 2012;7(2):e30966.

191. Petrov VV, Smirnova VV, Okorokov LA. Mercaptoethanol and dithiotreitol decrease the difference of electrochemical proton potential across the yeast plasma and vacuolar membrane and activate their H^+-ATPases. *Yeast* 1992;8:589–99.

192. Price NPJ, Ray KJ, Vermillion KE, Dunlap CA, Kurtzman CP. Structural characterization of novel sophorolipid biosurfactants from a newly identified species of *Candida yeast*. *Carbohydr Res* 2012;348:33–41.

193. Puchkov EO, Wiese A, Seydel U, Kulakovskaya TV. Cytoplasmic membrane of a sensitive yeast is a primary target for *Cryptococcus humicola* mycocidal compound (microcin). *Biochim Biophys Acta (Biomembranes)* 2001;1512:239–50.

194. Puchkov EO, Zahringer U, Lindner B, Kulakovskaya TV, Seydel U, Wiese A. Mycocidal, membrane-active complex of *Cryptococcus humicola*, is a new type of cellobiose lipid with detergent features. *Biochim Biophys Acta (Biomembranes)* 2002;1558:161–70.

195. Ratsep P, Shah V. Identification and quantification of sophorolipid analogs using ultra-fast liquid chromatography-mass spectrometry. *J Microbiol Methods* 2009;78:354–6.

196. Rau U, Manzke C, Wagner F. Influence of substrate supply on the production of sophorose lipids by *Candida bombicola*. ATTC 22214. *Biotechnol Lett* 1996;18:149–54.

197. Rau U, Nguyen LA, Schulz S, Wray V, Nimtz M, Roeper H, et al. Formation and analysis of mannosylerythritol lipids secreted by *Pseudozyma aphidis*. *Appl Microbiol Biotechnol* 2005;66:551–9.

198. Ribeiro IA, Bronze MR, Castro MF, Ribeiro MH. Sophorolipids: improvement of the selective production by *Starmerella bombicola* through the design of nutritional requirements. *Appl Microbiol Biotechnol* 2013;97:1875–87.

199. Rodrigues L, Banat IM, Teixeira J, Oliveira R. Biosurfactants: potential applications in medicine. *J Antimicrob Chemother* 2006;57:609–18.

200. Roelants SL, Saerens KM, Derycke T, Li B, Lin YC, Van de Peer Y, et al. *Candida bombicola* as a platform organism for the production of tailor-made biomolecules. *Biotechnol Bioeng* 2013;110:2494–503.

201. Rosenberg E, Ron EZ. High- and low-molecular-mass microbial surfactants. *Appl Microbiol Biootechnol* 1999;52:154–62.

202. Saerens K, Roelants S, Van Bogaert INA, Soetaert W. Identification of the UDP-glucosyltransferase gene UGTA1, responsible for the first glucosylation step in the sophorolipid biosynthetic pathway of *Candida bombicola* ATCC 22214. *FEMS Yeast Res* 2011;11:123–32.

203. Saerens KM, Zhang J, Saey L, Van Bogaert IN, Soetaert W. Cloning and functional characterization of the UDP-glucosyltransferase UgtB1 involved in sophorolipid production by *Candida bombicola* and creation of a glucolipid-producing yeast strain. *Yeast* 2011;28:279–92.

204. Saerens KM, Saey L, Soetaert W. One-step production of unacetylated sophorolipids by an acetyltransferase negative *Candida bombicola*. *Biotechnol Bioeng* 2011;108:2923–31.

205. Sandrin TR, Chech AM, Maier RM. A rhamnolipid biosurfactant reduces cadmium toxicity during naphthalene biodegradation. *Appl Environ Microbiol* 2000;66:4585–8.

206. Schippers C, Gessner K, Muller T, Scheper T. Microbial degradation of phenanthrene by addition of a sophorolipid mixture. *J Biotechnol* 2000;83:189–98.

207. Sen R, editor: Biosurfactants. Advance in Experimental Medicine and Biology. 2010. vol. 672. Springer. New York/Heidelberg.

208. Shah V, Doncel GF, Seyoum T, Eaton KM, Zalenskaya I, Hagver R, et al. Sophorolipids, microbial glycolipids with anti-human immunodeficiency virus and sperm-immobilizing activities. *Antimicrob Agents Chemother* 2005;49:4093–100.

209. Shah V, Jurjevic M, Badia D. Utilization of restaurant waste oil as a precursor for sophorolipid production. *Biotechnol Prog* 2007;23:512–5.

210. Shao L, Song X, Ma X, Li H, Qu Y. Bioactivities of sophorolipid with different structures against human esophageal cancer cells. *J Surg Res* 2012;173:286–91.

211. Shete AM, Wadhawa G, Banat IM, Chopade BA. Mapping of patents on bioemulsifier and biosurfactent: a review. *J Sci Industr Res* 2006;65:91–115.

212. Shibahara M, Zhao X, Wakamatsu Y, Nomura N, Nakahara T, Jin C, et al. Mannosylerythritol lipid increases levels of galactoceramide in and neurite out growth from PC12 pheochromocytoma cells. *Cytotechnology* 2000;33:247–51.

213. Sleiman JN, Kohlhoff SA, Roblin PM, Wallner S, Gross R, Hammerschlag MR, et al. Sophorolipids as antibacterial agents. *Ann Clin Lab Sci* 2009;39:60–3.

214. Soberón-Chávez G, editor: *Biosurfactants from genes to applications*. Series: Microbiology Monographs, 2011, 20, VIII, 216 p. Springer. New York / Heidelberg

215. Spencer JFT, Spencer DM, Tulloch AP. Extracellular glycolipids of yeasts. In: Rose AH, editor. *Economic microbiology, vol 3, secondary products of metabolism*. London, New York, San Francisco: Academic Press; 1979. p. 523–40.

216. Spoeckner S, Wray V, Nimtz M, Lang S. Glycolipids of the smut fungus *Ustilago maydis* from cultivation on renewable resources. *Appl Microbiol Boitechnol* 1999;51:33–9.

217. Stadler M, Bitzer J, Kopska B, Reinhardt K. Long chain glycolipids useful to avoid persisting or microbial contamination of materials, European Patent Application EP 2532 232 A1, Date of publication 12/12/ 2012.

218. Sun XX, Lee YJ, Choi JK, Kim EK. Synergistic effect of sophorolipid and loess combination in harmful algal blooms mitigation. *Mar Pollut Bull* 2004;48:863−72.

219. Susan TD, Hossack JA, Rose AH. Plasma-membrane lipid composition and ethanol tolerance in *Saccharomyces cerevisiae. Arch Microbiol* 1978;117:239−45.

220. Suzuki M, Kitagawa M, Yamamoto S, et al: Biosurfactant activators. mannosylerythritol lipid, and production methods, US Patent Application 2010, 20100168405.

221. Tabata N, Ohyama Y, Tomoda H, Abe T, Namikoshi M, Omura S. Structure elucidation of roselipins, inhibitors of diacylglycerol acyltransferase produced by *Gliocladium roseum* KF-1040. *J Antibiot (Tokyo)* 1999;52:815−26.

222. Takahashi M, Morita T, Wada K, Hirose N, Fukuoka T, Imura T, et al. Production of sophorolipid glycolipid biosurfactants from sugarcane molasses using *Starmerella bombicola* NBRC 10243. *J Oleo Sci* 2011;60:267−73.

223. Takahashi M, Morita T, Fukuoka T, Imura T, Kitamoto D. Glycolipid biosurfactants, mannosylerythritol lipids, show antioxidant and protective effects against H_2O_2-induced oxidative stress in cultured human skin fibroblasts. *J Oleo Sci* 2012;61:457−64.

224. Teichmann B, Linne U, Hewald S, Marahiel MA, Bölker M. A biosynthetic gene cluster for a secreted cellobiose lipid with antifungal activity from *Ustilago maydis. Mol Microbiol* 2007;66:525−33.

225. Teichmann B, Liu L, Schink KO, Bölker M. Activation of the ustilagic acid biosynthesis gene cluster in *Ustilago maydis* by the C2H2 zinc finger transcription factor Rua1. *Appl Environ Microbiol* 2010;76:2633−40.

226. Teichmann B, Lefebvre F, Labbe C, Bölker M, Linne U, Belanger RR. Beta hydroxylation of glycolipids from *Ustilago maydis* and *Pseudozyma flocculosa* by an NADPH-dependent β-hydroxylase. *Appl Environ Microbiol* 2011;77:7823−9.

227. Teichmann B, Labbe C, Lefebvre F, Bölker M, Linne U, Belanger RR. Identification of a biosynthesis gene cluster for flocculosin a cellobiose lipid produced by biocontrol agent *Pseudozyma flocculosa. Mol Microbiol* 2011;79:1483−95.

228. Thaniyavarn J, Chianguthai T, Sangvanich P, Roongsawang N, Washio K, Morikawa M, et al. Production of sophorolipid biosurfactant by *Pichia anomala. Biosci Biotechnol Biochem* 2008;72:2061−8.

229. Tomoda H, Ohyama Y, Abe T, Tabata N, Namikoshi M, Yamaguchi Y, et al. Roselipins, inhibitors of diacylglycerol acyltransferase, produced by *Gliocladium roseum* KF-1040. *J Antibiot (Tokyo)* 1999;52:689−94.

230. Trilisenko LV, Kulakovskaya EV, Kulakovskaya TV, AYu Ivanov, Penkov NV, Vagabov VM, et al. The antifungal effect of cellobiose lipid on the cells of *Saccharomyces cerevisiae* depends on carbon source. *SpringerPlus* 2012;1:18.

231. Tulloch AP, Spencer JF. Fermentation of long chain compounds by *Torulopsis magnoliae.* 3. Preparation of dicarboxylic acids from hydroxy fatty acid sophorosides. *J Am Oil Chem Soc* 1966;43:153−6.

232. Tulloch AP, Spencer JFT, Deinema MH. A new hydroxy fatty acid sophoroside from *Candida bogoriensis. Can J Chem* 1968;46:345−8.

233. Vagabov VM, Ivanov AY, Kulakovskaya TV, Kulakovskaya EV, Petrov VV, Kulaev IS. Efflux of potassium ions from cells and spheroplasts of *Saccharomyces cerevisiae* yeast treated with silver and copper ions. *Biochemistry (Moscow)* 2008;73:1224−7.

234. Van Bogaert INA, Saerens K, De Muynck C, Develter D, Soetaert W, Vandamme EJ. Microbial production and application of sophorolipids. *Appl Microbiol Biotechnol* 2007;76:23−34.

235. Van Bogaert INA, Develter D, Soetaert W, Vandamme EJ. Cloning and characterization of NADPH cytochrome P450 reductase gene (CPR) from *Candida bombicola*. *FEMS Yeast Res* 2007;7:922–8.

236. Van Bogaert INA, De Maeseneire SL, Develter D, Soetaert W, Vandamme EJ. Cloning and characterisation of the glyceraldehyde 3-phosphate dehydrogenase gene of *Candida bombicola* and use of its promoter. *J Ind Microbiol Biotechnol* 2008;35:1085–92.

237. Van Bogaert INA, De Maeseneire SL, Develter D, Soetaert W, Vandamme EJ. Development of a transformation and selection system for the glycolipid-producing yeast *Candida bombicola*. *Yeast* 2008;25:273–8.

238. Van Bogaert INA, Develter D, Soetaert W, Vandamme EJ. Cerulenin inhibits *de novo* sophorolipid synthesis of *Candida bombicola*. *Biotechnol Lett* 2008;30:1829–32.

239. Van Bogaert INA, Sabirova J, Develter D, Soetaert W, Vandamme EJ. Knocking out the MFE-2 gene of *Candida bombicola* leads to improved medium-chain sophorolipid production. *FEMS Yeast Res* 2009;9:610–7.

240. Van Bogaert INA, Demey M, Develter D, Soetaert W, Vandamme EJ. Importance of the cytochrome P450 monooxygenase CEP52 family for the sophorolipid-producing yeast *Candida bombicola*. *FEMS Yeast Res* 2009;9:87–94.

241. Van Bogaert INA, Roelants S, Develter D, Soetaert W. Sophorolipid production by *Candida bombicola* on oils with a special fatty acid composition and their consequences on cell viability. *Biotechnol Lett* 2010;32:1509–14.

242. Van Bogaert INA, Soetaert W. Sophorolipids. In: Soberón-Chávez G, editor. *Biosurfactants. Microbiology monographs*, **20**. Springer; 2011. p. 179–210.

243. Van Bogaert INA, Zhang J, Soetaert W. Microbial synthesis of sophorolipids. *Process Biochem* 2011;46:821–33.

244. Van Bogaert IN, Holvoet K, Roelants SL, Li B, Lin YC, Van de Peer Y, et al. The biosynthetic gene cluster for sophorolipids: a biotechnological interesting biosurfactant produced by *Starmerella bombicola*. *Mol Microbiol* 2013;88:501–9.

245. Vance-Harrop MH, de Gusmãoll NB, de Campos-Takaki GM. New bioemulsifiers produced by *Candida lipolytica* using D-glucose and babassu oil as carbon sources. *Brazilian J Microbiol* 2003;34:120–3.

246. Wadekar SD, Kale SB, Lali AM, Bhowmick DN, Pratap AP. Utilization of sweetwater as a cost-effective carbon source for sophorolipids production by *Starmerella bombicola* (ATCC 22214). *Prep Biochem Biotechnol* 2012;42:125–42.

247. Wakamatsu Y, Zhao X, Jin C, Day N, Shibahara M, Nomura N, et al. Mannosylerythritol lipid induces characteristics of neuronal differentiation in PC12 cells through an ERK-related signal cascade. *Eur J Biochem* 2001;268:374–83.

248. Walker-Caprioglio HM, Casey WM, Parks LW. *Saccharomyces cerevisiae* membrane sterol modifications in response to growth in the presence of ethanol. *Appl Environ Microbiol* 1990;56:2853–7.

249. Yamamoto S, Morita T, Fukuoka T, Imura T, Yanagidani S, Sogabe A, et al. The moisturizing effects of glycolipid biosurfactants, mannosylerythritol lipids, on human skin. *J Oleo Sci* 2012;61:407–12.

250. Yamamoto S, Fukuoka T, Imura T, Morita T, Yanagidani S, Kitamoto D, et al. Production of a novel mannosylerythritol lipid containing a hydroxy fatty acid from castor oil by *Pseudozyma tsukubaensis*. *J Oleo Sci* 2013;62:381–9.

251. Zhao X, Wakamatsu Y, Shibahara M, Nomura N, Geltinger C, Nakahara T, et al. Mannosylerythritol lipid is a potent inducer of apoptosis and differentiation of mouse melanoma cells in culture. *Cancer Res* 1999;59:482–6.

252. Zhao X, Murata T, Ohno S, Day N, Song J, Nomura N, et al. Protein kinase C alpha plays a critical role in mannosylerythritol lipid-induced differentiation of melanoma B16 cells. *J Biol Chem* 2001;276:39903–10.

253. Zhang L, Somasundaran P, Singh SK, Felse AP, Gross R. Synthesis and interfacial properties of sophorolipid derivatives. *Colloids Surf A Physicochem Eng Asp* 2004;240:75–82.

254. Zhang J, Saerens KM, Van Bogaert IN, Soetaert W. Vegetable oil enhances sophorolipid production by *Rhodotorula bogoriensis*. *Biotechnol Lett* 2011;33:2417–23.

255. Zhanga D, Spadaroa D, Garibaldia A, Gullinoa ML. Selection and evaluation of new antagonists for their efficacy against postharvest brown rot of peaches. *Postharvest Biol Technol* 2010;55:174–81.

256. Zhou QH, Kosaric N. Utilization of canola oil and lactose to produce biosurfactant with *Candida bombicola*. *J Am Oil Chem Soc* 1995;72:67–71.

257. Zhou S, Xu C, Wang J, Gao W, Akhverdiyeva R, Shah V, et al. Supramolecular assemblies of a naturally derived sophorolipid. *Langmuir* 2004;20:7926–32.

SUBJECT INDEX

Note: Page numbers followed by "*f*" and "*t*" refers to figures and tables respectively.

THE INDEX OF GENERIC NAMES